The Collapse Frequency of Structures

Dirk Proske

The Collapse Frequency of Structures

Bridges - Dams - Tunnels - Retaining Structures - Buildings

Springer

Prof. (FH) Dr.-Ing. habil. Dirk Proske, MSc.
Architektur, Holz und Bau
Berner Fachhochschule
Burgdorf, Bern, Switzerland

ISBN 978-3-030-97249-3 ISBN 978-3-030-97247-9 (eBook)
https://doi.org/10.1007/978-3-030-97247-9

Translation from the German language edition: Einsturzhäufigkeit von Bauwerken by Dirk Proske, © Der/die Herausgeber bzw. der/die Autor(en), exklusiv lizenziert durch Springer Fachmedien Wiesbaden GmbH, ein Teil von Springer Nature 2021. Published by Springer Fachmedien Wiesbaden. All Rights Reserved.

Responsible Editor: Ralf Harms
This Springer imprint is published by the registered company Springer Nature Switzerland AG
The registered company address is: Gewerbestrasse 11, 6330 Cham, Switzerland

Dedicated to my parents
Anneliese and Gerhard Proske

Acknowledgements

Parts of this work were carried out as part of a study for the Swiss Federal Railways (SBB). I would like to take this opportunity to thank the SBB and especially Herbert Friedl for their support.

Furthermore, some students have carried out work for the underlaying publications. These are, for example, Christof Hofmann, Lukas Heinzelmann and Michael Schmid. I would also like to thank them most sincerely.

Furthermore, I would also like to thank Prof. Panos Spyridis, with whom I was able to conduct the research for tunnels. He contributed significantly to the success of this study.

Additionally, I would like to thank the reviewers of the individual journal publications for their numerous comments and suggestions. They have contributed significantly to improving the quality of the journal papers and thus indirectly to this book.

Also, an intensive discussion on the topic of comparability of failure probabilities and collapse frequencies took place in autumn 2020 with Prof. Ton Vrouwenvelder, Prof. Jochen Köhler and Prof. Michael Havbro Faber. I thank them very much for their comments and support.

Finally, I thank my wife for her proofreading of the German version. This book is an extended translation of the German book.

Würenlingen
2021

Dirk Proske

Contents

Introduction and Initial Position 1

1.1 Introduction

Human life in and with buildings is an integral part of the daily experience of almost all people on earth. People practically live in a built environment. Figure 1.1 shows an image of the city of Tokyo as an example of this. The presence and use of a technical product on this scale is unique and indicates overwhelming benefits for humanity. According to the UN (United Nations 2018), every person on earth even has the right for housing.

In fact, the idea of protecting people from climatological influences, predators and attacks is one of the oldest inventions in human history, even before the invention of the wheel or the boat. The usefulness of protection against climatological influences can still be directly tested today: Mortality fluctuates over the year in the rhythm of the seasons (Proske 2022).

To be honest, however, one must admit that the idea of protection from climatological influences did not begin with humans: numerous animal species build dwellings, be they caves, nests or mounds. The linguistic connection to human structures is established by the German term “Bau” (construction), which is often used for animal burrows.

In addition to protecting the inhabitants from external influences, buildings also enable the movement of people and goods. At first glance, this seems contradictory. However, structures such as bridges, tunnels, dams and retaining structures are an essential part of the route of the means of transport such as rail or road. Along with the means of transport, they form the basis for implementing the necessary capacity. And this now exceeds the natural mass flows on earth (Proske 2022), because modern industrial societies are characterised by intensive traffic and by large mass, goods and energy transport.

D. Proske, *The Collapse Frequency of Structures*,
https://doi.org/10.1007/978-3-030-97247-9_1

Fig. 1.1 Image of Tokyo as an example of built environment. (Photo: *U. Proske*)

In addition, human buildings store extraordinarily large economic values. In Germany, real estate assets amounted to 14 trillion euros in 2016, around 80% of tangible assets (ZIA 2017) and more than three times the annual gross domestic product.

Buildings exist in very large numbers, but they are usually unique. Therefore, in addition to their economic value, they can also represent a significant cultural value. Mentioned here are only sacred buildings, historic centres of cities or unique castles.

Such sacred buildings also prove that structures can be extraordinarily durable. However, if necessary, buildings can also be designed and erected for short periods of use.

So far, human structures only exist on Earth. In the meantime, there are initial considerations for the creation of structures on the Moon or Mars (ESA 2016).

The gain in quality of life and safety through buildings imperatively requires sufficient safety in buildings. As historical collections of laws show, e.g. the Laws of Hammurabi, people were aware of this requirement early in human history and enshrined it in law (Fig. 1.2). This book explores the question of how safe structures actually are. Safety is understood here as collapse frequency and mortality. Both parameters are determined for different types of structures. The types of structures are bridges, dams, tunnels, retaining structures, buildings and general structures, stadiums and wind turbines.

Fig. 1.2 Stele of the Hammurabi Code (plaster cast in the Pergamon Museum Berlin of the original in the Louvre, approx. 2 000 years B.C.). (Photo: *D. Proske*)

However, the book begins with a discussion of the comparability of calculated failure probabilities, a safety format in civil engineering, and observed collapse frequencies.

The book is the result of numerous scientific publications on this topic (Proske 2016, 2017a, b, 2018a, b, c, 2019a, b, c; Proske et al. 2019; Proske 2020a, b, c, d; Proske et al. 2021; Hofmann et al. 2021; Proske and Schmid 2021; Spyridis and Proske 2021). In this respect, the book mainly summarises previous work and puts it into context.

The systematic investigation and evaluation of building collapses is also not new, see e.g. Scheer (2001). Thus, there are numerous individual case studies, but also collections of collapse investigations and even statistical evaluations for types of structures. There is even an *"International Journal of Forensic Engineering"*. However, a systematic investigation of the collapse frequency of all types of structures is still pending or has been carried out with the above-mentioned own studies. In the author's view, however, such an investigation is indispensable in order to evaluate the effectiveness and proportionality of modern safety concepts in the construction industry.

1.2 Aim of This Book

A safety concept is a method for achieving safety through appropriate measures. Safety is a state in which no further measures need to be implemented (Proske 2022). Measures can, for example, be of an organisational nature, such as the four-eyes principle, proof of competence of those involved in construction or the application of safety factors in structural analysis.

However, one has to check the success of the safety concept in practice. In the case of collapses and damage, a case-by-case examination is usually carried out within the framework of forensic investigations, in engineering mostly in the form of expert reports. Since buildings and structures, in contrast to various products in mechanical engineering, such as motor vehicles, are usually not mass-produced but unique, it makes sense to consider each individual case.

Nevertheless, sufficient safety must also be ensured over the sum of all structures. Or in other words: if a large number of structures collapse, each with an individual cause, a systematic fault cannot be ruled out that only becomes apparent when viewed as a whole. This was seen very clearly in the case of nuclear power plants. Every single major accident, such as Lucens, Three Mile Island, Chernobyl or Fukushima, was an individual event. Nevertheless, after each of these events, practically all nuclear power plants worldwide were re-examined and re-evaluated. Today we know that the safety requirements for first-generation nuclear power plants were clearly too low (Proske 2016, 2020a), but at the same time there was tremendous enthusiasm. For example, people wanted to install nuclear power plants directly in urban areas (Proske 2022).

The same question applies to buildings and structures: Do the safety concepts work and are the measures taken sufficient? This question can only be answered by looking at the sum of all structures. Since the current safety concept is probabilistic, it makes sense to record safety statistically across all structures, i.e. via the collapse frequency. Both, the probability of failure as a result of probabilistic calculation and the collapse frequency as a result of statistics, belong to the family of stochastics within mathematics (see Fig. 1.3).

In addition to the collapse frequency, mortality should also be taken into account. Unfortunately, the collapse frequency is only a zero-order risk parameter, as it only records what happens, not what the consequences are. Mortality is a real risk parameter because it takes into account the consequences. Other risk parameters are discussed when present and as needed for the individual types of structures.

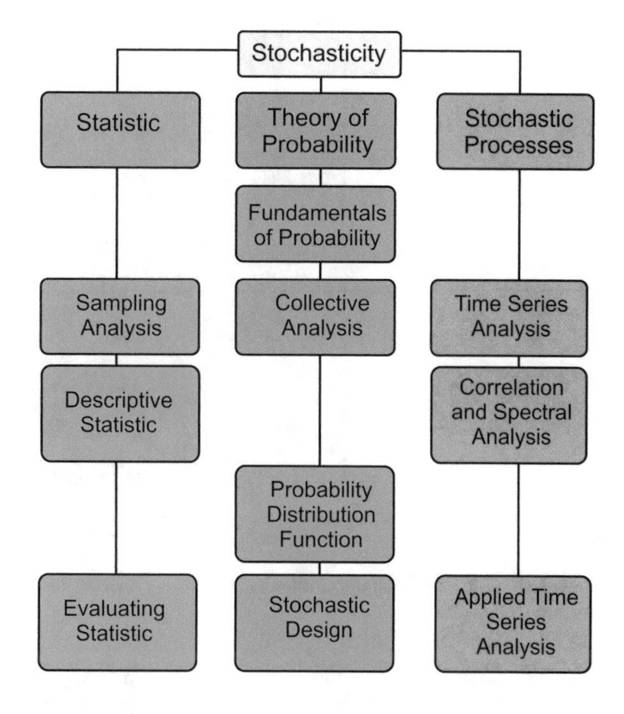

Fig. 1.3 Classification of the mathematical family of stochastics (Thoma 2004)

1.3 Comparability of Failure Probabilities and Collapse Frequencies

In the context of probabilistic and semi-probabilistic safety concepts, the so-called operational or nominal probability of failure is known. This indicates the probability of exceeding limit states based on certain models and input variables. The failure probability can refer to the ultimate limit state or the serviceability limit state. For the case of earthquakes, other limit states are known (FEMA 1997; Gkatzogias and Kappos 2015), which are closer to the actual collapse (see Fig. 1.4).

In addition, we know the observed collapse frequency, which is the subject of this book. This is calculated from the number of collapses in relation to the population of all structures for a reference area and a reference period.

Furthermore, there is the objective but unknown probability of collapse.

All three quantities, objective collapse probability, operational failure probability and observed collapse frequency are shown in a Venn diagram in Fig. 1.5.

The objective collapse probability includes all possible combinations of actions and resistances. It is complete in theory. In contrast, both the operational or nominal failure probability and the observed collapse frequency are incomplete. For example, the observation period of the collapses may have been too short, or the relevant actions may not have occurred. There may also be under- or overreporting, i.e. certain collapses were

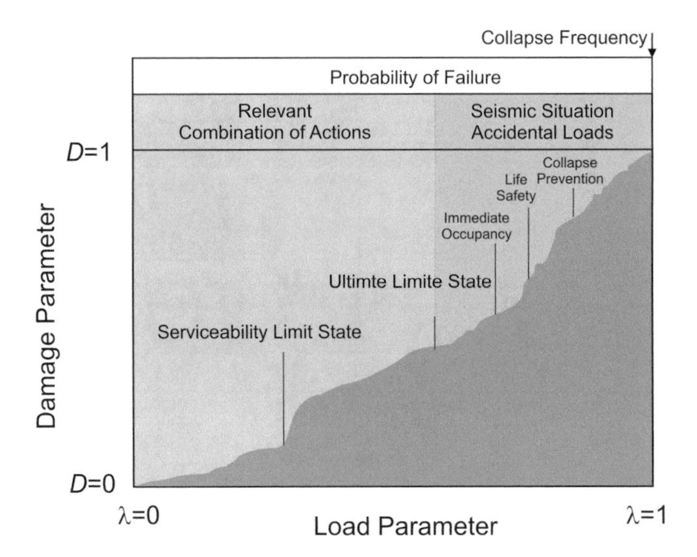

Fig. 1.4 Diagram load parameter vs. damage parameter including limit states

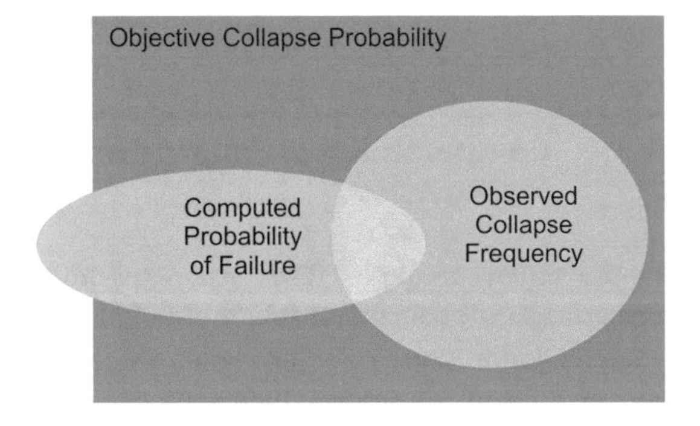

Fig. 1.5 Classification of the objective collapse probability, the operational or nominal failure probability and the observed collapse frequency

not documented. In the case of failure probabilities, we know that the relevant actions for bridges, such as flood or impact, were often not considered in the probabilistic investigations. In addition, human errors are simply excluded. Furthermore, the structural analysis equation in the ultimate limit state does not even have to be relevant for the load-bearing capacity. Deformation or crack width analysis can dominate and determine the design. However, this is usually not taken into account in the operational failure probability analysis. Figure 1.6 shows possible correction factors for the conversion of failure probabilities into collapse frequencies (Proske et al. 2020). As the figure shows, some factors reach considerable amounts.

The principal differences between failure probabilities and collapse frequencies are briefly mentioned below:

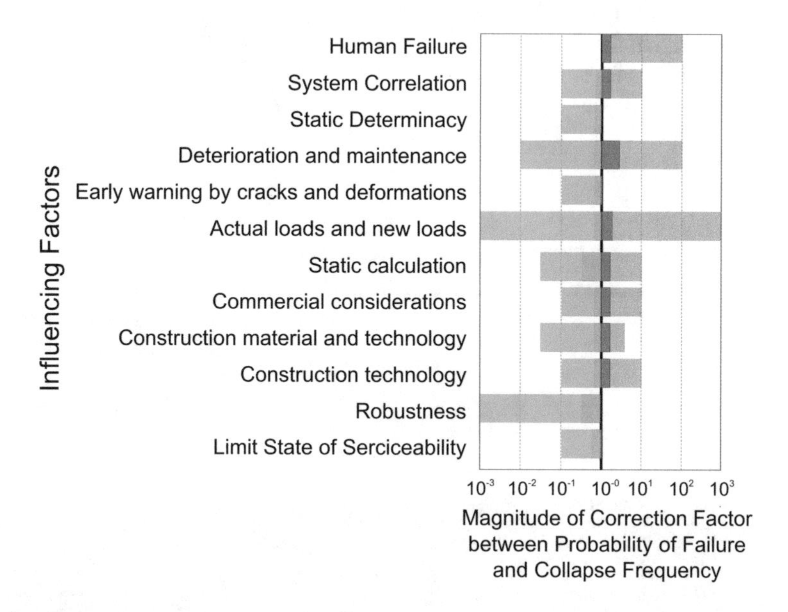

Fig. 1.6 Possible factors for converting the probability of failure into a collapse frequency (Proske et al. 2020)

- The calculated failure probabilities refer to individual components, individual structural analysis equations and individual structures, while the collapse frequencies refer to the population of all structures.
- The concept of failure probabilities is a bottom-up approach. Starting from individual parameters, such as concrete strength or geometry values, the behaviour of the component and, based on this, of the structure is modelled. In contrast, with collapse frequencies only the overall result in terms of a defined collapse is considered. The behaviour of the individual components or structural analysis is not of interest. This corresponds to a top-down approach.
- The failure probabilities are calculated using clearly defined limit states based on e.g. forces, strains, stresses or crack widths. The definition of collapse of structures in the calculation of collapse frequencies is not based on these parameters, but on a summary parameter, such as irreparable damage according to the definition of collapse.
- The calculated failure probabilities refer to future events, while collapse frequencies refer to events in the past.
- The calculated failure probability could possibly converge against the collapse frequency if the structures were the same. However, the structures are individual (see Fig. 1.7).
- Existing structures are usually subject to maintenance management with monitoring and repair. This can prevent the collapse of structures, as e.g. Figure 1.8 shows.

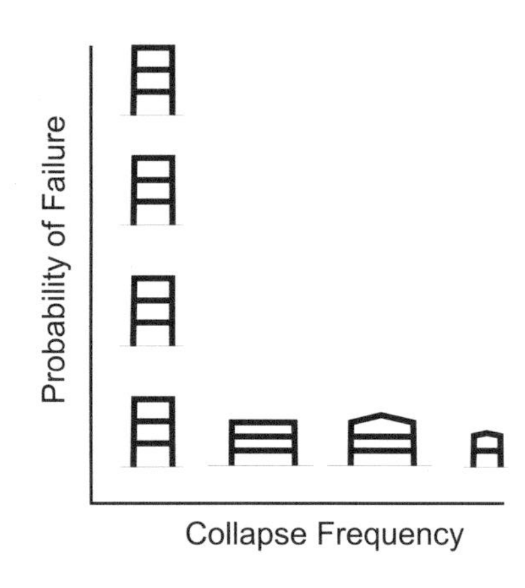

Fig. 1.7 Comparison of the failure probability and collapse frequencies with regard to the structure types

Fig. 1.8 Precautionary closure of a bridge in Germany. (Photo: *D. Proske*)

- The theoretical proof of the completeness of the actions cannot be provided when calculating failure probabilities. In the observed collapse frequency, completeness depends on chance. While the failure probability excludes human errors by definition, these are automatically included in the collapse frequency. However, the collapse frequency also takes positive influences into account. Such influences can be, for example, material over-strengths or over-dimensioning in the final state due to design-relevant actions during the construction state.

- Neither can the completeness of the actions be provided when calculating the failure probabilities, nor can the completeness of the stock and the collapses be provided when determining the collapse frequencies. Frequently, under-reporting (collapses are not recorded) or over-reporting (collapses are recorded several times) occurs.
- The collapse frequency can be influenced by reference areas or reference times that are too small. For example, it is known that significantly fewer large floods occurred on the Rhine in Switzerland in the twentieth century than in the centuries before (Wetter et al. 2011). The same effect has been discussed for storms (Usbeck et al. 2010). The collapse frequency may show much too good values if the relevant action did not occur in the observation period.
- It is common knowledge that engineers include conservative assumptions in their calculations. These are often difficult to quantify. An example of this is the requirement of reinforcement for unintended restraint in reinforced concrete slabs or the neglect of load-bearing elements in the structural analysis, such as backfill in arch bridges (see e.g. Proske et al. 2021). These conservative assumptions can be made at least partially visible through experimental load tests. They are currently referred to as "hidden safety" and are not part of the calculation of the failure probabilities. However, such effects are included in the collapse frequencies.
- The comparability of calculated core damage probabilities and observed core damage frequencies is the subject of intensive discussion in the field of nuclear engineering (Kauermann and Küchenhoff 2011; Raju 2016; de Vasconcelos et al. 2015; Wheatley et al. 2015; Proske 2016).
- The comparability of calculated probabilistic earthquake hazards and observed earthquake frequencies is the subject of intense discussion in the field of seismic hazard analysis (Stein et al. 2015; Kossobokov and Nekrasova 2012; Musson 2012; Stirling 2012; Wang 2008). Especially after the completion of the PEGASOS study, this discussion gained intensity (Klügel 2005; 2012; Aspinall 2010). A quote from one such discussion contribution: *"If one discards the idea that the reliability of PSH (probabilistic seismic hazard) estimates cannot be evaluated (this is equivalent to state that PSH assessment is not a scientific procedure), the most obvious (and frequent) position is to presume that the most recent estimates, presumably based on upgraded data sets, are the best ones. However, this position is unsatisfactory. As an example, it is not easy to elaborate shared opinions about the actual improvements induced by new data and hypotheses,"*
- In fact, such comparisons of earthquake frequency with a return period of 10,000 years (Safe Shutdown Earthquakes: SSE) have been made for nuclear power plants. Including 2011, only one or two units should have been exposed to such an earthquake, but in fact there were more than 20 (see Table 1.1).

Table 1.1 Number of nuclear power plants exposed to an SSE

Nuclear Power Plant	Country	Year	Number of Units	Remark
Perry	US	1986	1	Before licensing
Metsamor	Armenia	1989	2	
Kashiwazaki-Kariwa	Japan	2007	7	Exceedance>2
Fukushima-Daiichi	Japan	2011	6	Minor Exceedance
Onagawa	Japan	2011	3	
North-Anna	US	2011	2	
Overall			21	

The two quantities therefore differ substantially in calculation and interpretation. For this reason, numerous publications and standards explicitly assume that operational failure probabilities and observed collapse frequencies are not comparable and must not be mixed (Bolotin 1969; Kulhawy et al. 1983; Spaethe 1992; Melchers 1999; Ellingwood 2001; Imhof 2004; Oberguggenberger and Fellin 2005; Vogel et al. 2009; Eurocode 0 2017; FWF 2018, 2019; Moan and Eidem 2020). Three examples of incomparability are:

- Oberguggenberger and Fellin (2005) state: *"In particular, contrary to common language, the failure probability cannot be interpreted as a frequency of failure. This fact was already pointed out by Bolotin as early as 1969."*
- FWF (2019) states: *"Although not being the real frequency of collapse, the target value (probability of failure) can still be used …"*
- In Eurocode 0 (2017), section C 3.4 Reliability requirements states in Note 1: *"The specified reliability requirements governing the ultimate limit state and serviceability limit state design do not consider human errors. Therefore, failure probabilities are not directly related to observed failure rates, which are strongly influenced by failure modes that include some effects of human error."*

Two examples of the different values are:

- Spaethe (1992) states: *"[the operational failure probability contains] only a share of the total failure probability … possible shares from human error [are] not included in this theoretical value. If one can assume the mechanical model to be error-free, then the failure frequency will be greater than the theoretical failure probability."*
- FWF (2018) states: *"theoretical probability of failure is orders of magnitude lower than the actual frequency of failures … actual failures are due to causes that are beyond theoretical probabilities…"*

Despite the differences, calculated probabilities and observed frequencies are often used and mixed as equivalent in real decisions. This is demonstrated by the following thirteen examples:

- According to Eurocode 0 (2017), design values for building materials can be determined both from experience and by probabilistic calculations (see Fig. 1.9).
- In life cycle cost analyses, calculated failure probabilities are used for the evaluation of the real building stock (Enright et al. 1999; Estes and Frangopol 2001; Li 2004; Neumann et al. 2010; Braml 2010; Davis-Mcdaniel 2011; Ghodoosipoor 2013; CEN 2013; Zanini et al. 2017; Proske 2018c; Strauss et al. 2018, 2019). The probability of failure is mostly related to the observed condition ratings of the structures. In Proske et al. (2020), collapse frequencies determined via adjustment factors from condition-based failure probabilities were used for the first time.
- Kleijn (2019) sees the calculated failure probabilities and the observed collapse frequencies as an equivalent basis for risk assessments of buildings (Sect. 3.4.1).
- Knoll (1965) writes *"One can also regard the structures in their entirety as a population in the statistical sense that can be observed. Without taking specific circumstances into account, one finds in all structures that have existed or still exist a qualification that is closely related to safety, namely the relative frequency of failures among the load-bearing structures."*
- Holicky and Sykora (2009) perform a limited comparison of observed snow-related roof collapses for the winter 2005/2006 with calculated failure probabilities.

Section 8.3.6 Design values of material properties

...

(2) Provided that the level of reliability is no less than that implied by the use of Formula (8.23), the design value of a material property may be determined directly from:

- empirical or theoretical relations with measured physical properties;
- physical and chemical composition;
- **from previous experience**;
- in geotechnical design, prescriptive measures;
- in geotechnical design, the most unfavourable value that the parameter could practically adopt;
- values given in European Standards or other documents that are specified in the other Eurocodes;
- **reliability analysis,** see Annex C; or
- results of tests, see Annex D.

Fig. 1.9 Extract from Eurocode 0 (2017) for determining design values

- Probabilistic models are fitted within the framework of experimental structural tests (Casas and Gomez 2013; Schmidt et al. 2020; De Koker et al. 2020; Proske et al. 2021).
- For existing structures, input variables for the probabilistic calculations on the structure are likely based on samples taken from the structure (Eurocode 0 2017). From these samples, probability distributions are constructed via frequency distributions and statistical estimators, and the probability distributions are used in the probabilistic calculations.
- Nuclear power plant strengthening, and refurbishment are based at least in part on probabilistic calculations (Richner et al. 2010; ENSI 2018).
- In many probabilistic calculations, probabilistic input variables are adjusted by observations using Bayesian update (Eurocode 0 2017; ENSI 2018). In the field of nuclear energy, for example, generic data are linked with observed failure data (AXPO, Stetkar & Associates, ABS Consulting 2013).
- Earthquake evidence is at least partly based on observations (Wenk 2005; Kölz and Duvernay 2005 see Fig. 1.10). Iervolino et al. (2018) use a Global Collapse Failure Rate as a calculated system failure probability taking into account the return period of real acceleration time histories.
- The definition of seismic fragilities (conditional failure probabilities depending on earthquake intensity) was based at least partly on observations (EPRI 1994; Yanev and Yanev 2013).
- The setting of probabilistic fire safety targets is based at least in part on observed mortalities (VKF 2018)
- The collapse frequency of structures during fires is the basis for the development of action instructions for firefighters, since such collapses often lead to the death of the emergency workers (NIOSH 1999; Stroup and Bryner 2007).

In fact, one cannot exclude the comparison because the calculated probabilities and observed frequencies are mixed in calculations and because they are both used equally for decisions.

In addition, of course, the probabilistic calculations have an impact on the observed values if, for example, reinforcement measures result from the calculations, and the observed values have an impact on the probabilistic calculations if, for example, a severe accident leads to regulations being tightened and more reinforcements having to be implemented. The latter is often referred to in the literature as *"design by disaster"* (De Sanctis 2015, see also Birdgle and Sims 2009; Ratay 2010, 2011).

```
2.3 Recording the Probability of Collapse
2.3.1 General

The collapse probability is understood here as the probability of a building
collapsing more or less completely as a result of an earthquake of a certain
magnitude. The following ideas apply as basic values:
- The collapse probability of a certain building standing in earthquake zone 1 is
  10 times lower than that of a building in earthquake zone 3b.
- The probability of collapse of one and the same building on good ground is
  50% less, and twice as much on poor ground as on average ground.
- The collapse probability of a certain building designed according to the
  SIA 160 (1989) standard is the same in all seismic zones.
- The collapse probability of one and the same building designed before 1970 is
  three times greater in earthquake zone 1 than that of an analogous building
  designed for earthquakes after 1989 in accordance with SIA 160.
  For zone 3b a factor of 15 applies here.
- The probability of collapse of a building that is favourably designed and
  seismically designed for earthquakes is about 20 times smaller than that of a
  building that is poorly designed in almost every respect.
As a result, it is not the probability itself that is determined as an absolute
value but rather a key figure WZ is determined. This can be used to compare
buildings with each other. The collapse probability is recorded using indicators,
to which corresponding key figures are assigned. These key figures do not quantify
the probability of collapse in an absolute manner, but are coordinated auxiliary
variables that serve to classify the collapse probability. ...

2.3.4 Collapse Probability Index

The key figure for the collapse probability WZ summarises all the above-mentioned
characteristics. It is obtained by processing the key figures determined for the
structure according to the following formula::

WZ = WEP WB (1 + WG + WA + WW + WK + WD + WF)
```

Fig. 1.10 Computation of the seismic collapse probability of structures according to Kölz and Duvernay (2005)

1.4 Theoretical Considerations

According to the famous sentence *"All models are wrong, some are useful"*, one must basically assume a limited comparability of model results and observations. Models are never objective (independent of the observer) and true (one can only approach truth asymptotically, Bavink 1944). Models can even contradict each other and only explain individual facets of observations, but not the overall observation. This is known, for example, from physics with the duality of light as particle and wave, but also in the field of risk assessments (Nielsen et al. 2019). In this context, reference should be made to a quotation by Pestalozzi *"It is the lot of men that no one has the truth. They all have it but distributed."* One must be aware of these limitations of the models when applying them.

Figure 1.11 shows an example of the development of a model within the whole environment. As can be seen in the figures, different possibilities exist to cut the system or model free and to simplify the interactions with the environment. To improve the development of the models for engineers and scientists, there are subject-specific recommendations, e.g. in standards or guidelines, or some general recommendations, e.g.

- A calculation model should be convergent, i.e. the more accurate the input variables become, the more accurate the result should become.
- A calculation model should be robust, i.e. small changes in the input variables should lead to small changes in the calculation results.
- A calculation model should have no systematic error, i.e. the mean statistical error should be zero.

Furthermore, the following should apply

- The input variables of the calculation model must be measurable or calculable. (Furthermore, in the calculation, highly weighted input variables should be more accurately determinable than low weighted input variables).
- A calculation model should be easy to handle and practical.
- A calculation model should have at least a rudimentary theoretical background.
- A calculation model should be compliant with historical calculation models in the areas that have proven themselves.

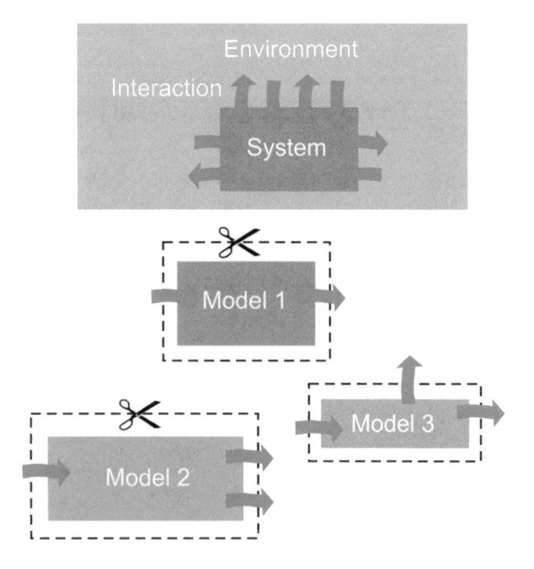

Fig. 1.11 Different ways to cut a model free

- If two calculation models have the same accuracy, one should choose the model that requires less input data.

The choice of relevant parameters is sometimes context-dependent. This is shown in Fig. 1.12. The distance to the centre is supposed to represent the objective importance of the input variables to the question. In fact, however, the factors are not selected according to this, but according to a variety of other reasons, such as availability (availability heuristic).

Due to their incompleteness, models must always be tested for the fulfilment of their purpose. This can only be done through interaction with reality. In the engineering sciences this is usually done through experiments, in the social sciences through surveys. It is sometimes called the Wheel of Science (Wallace 1971, see Fig. 1.13).

However, even with these tests, errors and deviations can occur. This is why there is also the saying: *"If you measure, you measure crap"*. For this reason, Fig. 1.13 also shows unconsidered influences in the measurements. Figure 1.13 shows three cases. In case a) a model is developed which is not used in reality. This means that the model cannot be tested. In case b) the model is applied in reality, but there is no check of the model, i.e. no feedback of information from observations to the model. In case c), the model is both applied in reality and tested. The determination of the collapse frequencies of structures is a contribution to case c).

Furthermore, it is imperative to check the calculated probabilities of failure against a quantity that can be observed in reality. Otherwise, the calculation of the risk, which is itself a theoretical construct, is also based only on theoretical constructs. This is dangerous in its consequence because buildings and structures are real and not a theoretical construct.

The regular and frequent testing of engineering models is indispensable in the empirical sciences. Otherwise, the result is something like in mathematics, where proofs can be

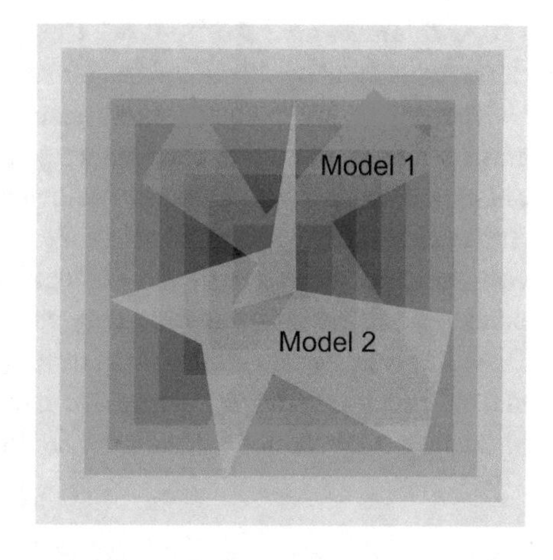

Fig. 1.12 Selection of relevant model parameters in two cases (A and B). The actual relevance of the parameters is represented by the shades of grey: the darker, the more relevant

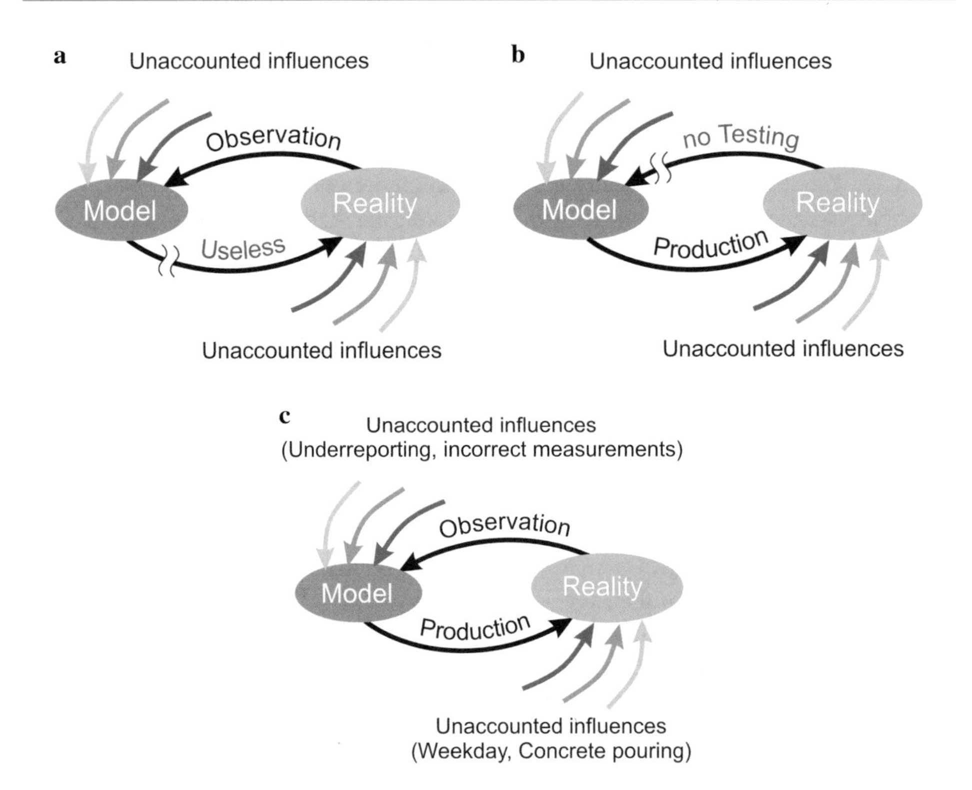

Fig. 1.13 Model-Reality Cycle (adaption of the Wheel of Science)

produced, but these do not necessarily have to have a connection to reality (Fig. 1.13 a and b respectively).

Comparing what scientists and engineers calculate with what they observe in reality is an essential part of scientific work (Popper 1993). As Albert Einstein observed, *"All essential ideas in science are born out of the dramatic conflict between reality and our efforts to understand that reality."* Comparison is thus indispensable for the successful development and application of engineering and technology.

The use of probabilistic safety concepts for the vast majority of new buildings on earth while at the same time arguing that the calculated failure probabilities should not be compared with collapse frequencies, as put forward e.g. in Spaethe (1992) or Eurocode 0 (2017), seems questionable. This does not mean that failure probabilities and collapse frequencies are directly comparable. Inevitably, however, there must be a connection when the probabilistic safety concept is applied to real structures. Either the collapse frequencies remain the same or they change. A control of the failure probabilities, and this also includes the normative target values, must therefore be mandatory. Even alternative methods for determining the target values, such as the application of the Life

Quality Index, require a check of the results in reality. They are therefore no substitute for the test carried out here.

References

Aspinall W (2010) A route to more tractable expert advice. Nature 463:294–295

AXPO, Stetkar & Associates, ABS Consulting (2013) Beznau Risk Assessment (BERA) – Power Operation and Hot Shutdown SG Cooling, December 2013

Bavink B (1944) Ergebnisse und Probleme der Naturwissenschaften, Achte. Verlage von S, Hirzel, Leipzig

Birdgle RJ, Sims FA (2009) The effect of bridge failures on UK technical policy and practice. Proceedings of the Institution of Civil Engineers, Engineering History and Heritage 162(EH1):39–49. https://doi.org/10.1680/ehh2009.162.1.39. Accessed 4. July 2019

Bolotin VV (1969) Statistical methods in structural mechanics. Holden-Day Inc., San Francisco

Braml T (2010) Zur Beurteilung der Zuverlässigkeit von Massivbrücken auf der Grundlage der Ergebnisse von Überprüfungen am Bauwerk, Dissertation, Universität der Bundeswehr München, München

Casas RJ, Gomez JD (2013) Load rating of highway bridges by proof-loading, April 2013. KSCE J Civ Eng 17(3):556–567

CEN (2013) CEN Workshop Agreement: Ageing behavior of Structural Components with regard to Integrated Lifetime Assessment and subsequent Asset Management of Constructed Facilities, CWA 16633, May 2013

Davis-Mcdaniel C (2011) Fault-tree model for bridge collapse risk analysis. Thesis, Clemson University, M.Sc

De Koker N, Vilkoen C, Lenner R, Jacobsz SW (2020) Updating structural reliability efficiently using load measurement. Struct Saf 84:1–9 (May 2020, 101939)

De Sanctis G (2015) Generic Risk Assessment for Fire Safety: Performance Evaluation and Optimisation of Design Provisions, Doctoral Thesis, IBK Bericht, Vol 363. Institut für Baustatik und Konstruktion der ETH Zürich, Zürich

de Vasconcelos V, Soares WA, da Costa ACL (2015) FN-Curves: Preliminary Estimation of Severe Accident Risks After Fukushima. 2015 International Nuclear Atlantic Conference (INAC 2015), Sao Paulo, Brasilien, 4.–9. Oktober 2015, Associacao Brasileira De Energia Nuclear (ABEN), 12 pages

Ellingwood BR (2001) Acceptable risk bases for design of structures. Prog Struct Mat Eng 3(2):170–179

Enright MP, Frangopol DM, Gharaibeh ES (1999) Reliability of Bridges under Aggressive Conditions. Application of Statistics and Probability (ICASP 8). Sydney 1:323–330

ENSI (2018) Probabilistische Sicherheitsanalyse (PSA): Qualität und Umfang, Richtlinie für die schweizerischen Kernanlagen, ENSI-A05, Eidgenössisches Nuklearsicherheitsinspektorat, Ausgabe Januar 2018, Brugg.

EPRI (1994) Methodology for developing seismic fragilities, prepared by J R Benjamin and Associates, Inc and RPK Structural Mechanics Consulting, TR-103959, Project 2722–23

ESA (2016) Jan Wörner: Moon Village – Menschen und Roboter gemeinsam auf dem Mond, https://www.esa.int/Space_in_Member_States/Germany/Jan_Woerner_Moon_Village_-_Menschen_und_Roboter_gemeinsam_auf_dem_Mond

Estes AC, Frangopol DM (2001) Bridge lifetime system reliability under multiple limit states. J Bridge Eng, ASCE 6(6):523–528

Eurocode 0 (2017) EN 1990 Basis of structural design, 2nd edn, Draft 30 April 2017
FEMA (1997) FEMA 273 - NEHRP Guidelines for the seismic rehabilitation of buildings, Federal Emergency Management Agency, October 1997, Washington D.C., Redwood City
FWF (2018) Austrian Science Fund, Expert Report: Comparison of probability of failure and frequency of collapse, 2. Juli 2018, Wien
FWF (2019) Austrian Science Fund, Expert Report: Comparison of probability of failure and frequency of collapse, 20 März 2019, Wien
Ghodoosipoor F (2013) Development of Deterioration Models for Bridge Decks using System Reliability Analysis, Dissertation, Concordia University, Montreal, Quebec, Canada
Gkatzogias KI, Kappos AJ (2015) Deformation-based seismic design of concrete bridges. Earthquakes and Struct 9(5):1045–1067
Hofmann C, Proske D, Zeck K (2021) Vergleich der Einsturzhäufigkeit und Versagenswahrscheinlichkeit von Stützbauwerken. Bautechnik 98(7):475–481.
Holicky M, Sykora M (2009) Failures of Roofs under Snow Load: Causes and Reliability Analysis, Fifth Forensic Engineering Congress, November 11–14 2009, Washington DC, 10 pages
Iervolino I, Spillatura A, Bazzurro P (2018) Seismic reliability of code-conforming Italian buildings. J Earthquake Eng 22(S2):5–27
Imhof, D. (2004) Risk Assessment of Existing Bridge Structures. Cambridge, University of Cambridge, Dissertation
Kauermann G, Küchenhoff H (2011) Reaktorsicherheit: Nach Fukushima stellt sich die Risikofrage neu, Frankfurter Allgemeine Zeitung, 2011. https://www.faz.net/aktuell/politik/energiepolitik/reaktorsicherheit-nach-fukushima-stellt-sich-die-risikofrage-neu-1605610.html. Accessed 3. Dec. 2019
Kleijn SMJ (2019) Study for a Quantitative Risk Assessment on Accidental Actions, MSc Thesis, TU Delft, Delft
Klügel J-U (2005) Problems in the application of the SSHAC probability method for assessing earthquake hazards at Swiss nuclear power plants, Eng Geol 78(3/4):285–307
Klügel J-U (2012) Comment on "Earthquake Hazard Maps and Objective Testing: The Hazard Mappers Point of View" by Mark W. Stirling. Seismol Res Lett 83(5):829–830
Knoll F (1965) Grundsätzliches zur Sicherheit der Tragwerke, Promotion 3701, ETH Zürich, Zürich
Kölz E, Duvernay B (2005) Beurteilung der Erdbebensicherheit bestehender Gebäude, Konzept und Richtlinien für die Stufe 1 (2. Fassung), Bundesamt für Wasser und Geologie, BWG, Biel
Kossobokov VG, Nekrasova AK (2012) Global Seismic Hazard Assessment Program Maps are Erroneous. Seismic Instruments 48(2):162–170
Kulhawy FH, Trautmann CH, Beech JF, O'Rourke TD, McGuire M (1983) Transmission line structure foundations for uplift-compression loading. Technical Report, Electric Power Research Institute
Li Y (2004) Effect of Spatial Variability on Maintenance and Repair Decisions for Concrete Structures, Promotion, TU Delft, 2004. https://pdfs.semanticscholar.org/7375/cbc6eb7ad587cbb20bf14d3be34673f97aac.pdf
Melchers RE (1999) Structural reliability analysis and prediction. John Wiley & Son Ltd.
Moan T, Eidem ME (2020) Floating bridges and submerged tunnels in Norway – history and the future outlook, In: Wang CM, Lim SH, Tay ZX (eds): WCFS2019: Proceedings of the World Conference on Floating Solutions, Singapore, pp. 81–111
Musson RMW (2012) PSHA validated by quasi observational means. Seismol Res Lett 83(1):130–134

Neumann T, Vogel M, Müller HS (2010) Quantification of the Durability of Concrete Bridges – a Probabilistic Modelling of Environmental and Functional Interactions, P. van Gelder, L. Gucma, D. Proske: Proceedings of the 8th International Probabilistic Workshop, Szczecin, pp. 261–274

Nielsen L, Glavind ST, Qin J, Faber MH (2019) Faith and fakes – dealing with critical information in decision analysis. Civ Eng Environ Syst 36(1):32–54

NIOSH (1999) NIOSH Alert: Preventing Injuries and Deaths of Fire Fighters due to Structural Collapse, U.S. Department of Health and Human Services (DHHS), National Institute for Occupational Safety and Health (NIOSH) Publication No. 99–110, August 1999

Oberguggenberger M, Fellin W (2005) The fuzziness and sensitivity of failure probabilities. In: Fellin W, Lessmann H, Oberguggenberger M, Vieider R. (eds) Analysing Uncertainty in Civil Engineering. 2005, Springer, Berlin

Popper KR (1993) Alles Leben ist Problemlösen – Über Erkenntnis, Geschichte und Politik. Piper, München

Proske D (2016) Differences between probability of failure and probability of core damage. In: Caspeele R, Taerwe L, Proske D (eds) Proceedings of the 14th International Probabilistic Workshop, Ghent, Springer, pp. 109–122

Proske D (2017) Vergleich der Versagenswahrscheinlichkeit und der Versagenshäufigkeit von Brücken. Bautechnik 94(7):419–429

Proske D (2017b) Comparison of Bridge Collapse Frequencies with Failure Probabilities, Proceedings of the 15th International Probabilistic Workshop, Dresden, TUDpress, Eds. M. Voigt, D. Proske, W. Graf, M. Beer, U. Häussler-Combe, P. Voigt, September 2017, pp. 15–23

Proske D (2018a) Versagenshäufigkeit und Versagenswahrscheinlichkeit von Brücken, Tagungsband des 28. Technische Universität Dresden, Dresden, Dresdner Brückenbausymposium, pp 189–199

Proske D (2018b) Comparison of Large Dam Failure Frequencies with Failure Probabilities, Beton- und Stahlbetonbau, Vol 113, Sonderheft (S2): 16th International Probabilistic Workshop, pp. 2–6

Proske D (2018c) Bridge Collapse Frequencies versus Failure Probabilities. Springer, Cham

Proske D (2019a) Comparison of the collapse frequency and the probability of failure of bridges. Proc Inst Civ Eng – Bridge Eng 172(1):27–40

Proske D (2019b) Ist der Vergleich von Einsturzhäufigkeiten und Versagenswahrscheinlichkeiten sinnvoll? ce papers 3(2):48–53

Proske D (2019c) Comparison of Frequencies and Probabilities of Failure in Engineering Sciences, Proceedings of the 29th European Safety and Reliability Conference, edited by M Beer and E Zio, Research Publishing, Singapore, pp. 2040–2044. https://doi.org/10.3850/978-981-11-2724-30816-cd

Proske D (2020a) Erweiterter Vergleich der Versagenswahrscheinlichkeit und -häufigkeit von Kernkraftwerken, Brücken. Dämmen Und Tunneln. Bauingenieur 95(9):308–317

Proske D (2020b) Zur Berücksichtigung hypothetischer Opferzahlen in Lebenszykluskostenberechnungen von Brücken. Beton- und Stahlbetonbau 115(6):459–546

Proske D (2020c) Fatalities due to Bridge Collapse. Proc Inst Civ Eng – Bridge Eng 173(4):260–267

Proske D (2020d) Die globale Gesundheitsbelastung durch Bauwerksversagen. Bautechnik 97(4):233–242

Proske D (2022) Katalog der Risiken, 2. vollständig überarbeite Aufl. Springer, Berlin, Heidelberg

Proske D, Hostettler S, Friedl H (2020) Korrekturfaktoren der Einsturzwahrscheinlichkeit von Brücken. Beton- und Stahlbetonbau 115(2):128–135

Proske D, Schmid M (2021) Häufigkeit von und Mortalität bei Hochbaueinstürzen. Bautechnik 98(6):423–432.

Proske D, Spyridis P, Heinzelmann L (2019) Comparison of Tunnel Failure Frequencies and Failure Probabilities. Yurchenko D & Proske D (eds): Proceedings of the 17th International Probabilistic Workshop, Edinburgh, pp. 177–182

Proske D, Sykora M, Gutermann M (2021) Verringerung der Versagenswahrscheinlichkeit von Brücken durch experimentelle Traglastversuche. Bautechnik 98(2):80–92

Raju S (2016) Estimating the frequency of nuclear accidents. Science & Global Security 24(1):37–62

Ratay RT (2010) Changes in Codes, Standards and Practice Following Structural Failures, Part 1: Bridges, Structure magazine, December 2010, pp. 16–19

Ratay RT (2011) Changes in Codes, Standards and Practice Following Structural Failures, Part 2: Buildings, Structure magazine, April 2011, pp. 21–24

Richner M, Tinic S, Proske D, Ravindra M, Campbell R, Beigi F, Asfura A (2010) Insights gained from the Beznau Seismic PSA, Proceedings of the 10th International Probabilistic Safety Assessment & Management Conference, June 2010, Seattle, IABSAM, auf CD, 11 pages

Scheer J (2001) Versagen von Bauwerken, Band 2, Hochbauten und Sonderbauwerke, Ernst und Sohn, September 2001, Berlin

Schmidt JW, Thöns S, Kapoor M (2020) Christensen CO, Engelund S, Sörensen JS: Challenges Related to Probabilistic Decision Analysis for Bridge Testing and Reclassification, Front. Built Environ 28(Februar):1–14

Spaethe G (1992) Die Sicherheit tragender Baukonstruktionen, zweite, neubearbeite. Springer, Wien

Spyridis P, Proske D (2021) Revised Comparison of Tunnel Collapse Frequencies and Tunnel Failure Probabilities. ASCE-ASME J Risk Uncertainty Eng Syst, Part A: Civ Eng 7(2):04021004-1–04021004-9

Stein S, Spencer BD, Brooks EM (2015) Metrics for assessing earthquake-Hazard map performance. Bull Seismol Soc Am 105(4):1–14

Stirling MW (2012) Earthquake hazard maps and objective testing: The hazard mapper's point of view. Seismol Res Lett 83(2):231–232

Strauss A, Lener G, Schmid J, Matos J, Casas JR (2018) Lebenszyklus- und Qualitätsspezifikationen für Ingenieurbauwerke, 28. Dresdner Brückenbausymposium, TU Dresden, Dresden, pp. 169–186

Strauss A, Mold L, Bergmeister K, Mandic A, Matos JC, Casas JR (2019) Performance Based Design and Assessment – Level of Indicators, Life-Cycle Analysis and Assessment in Civil Engineering: Towards an Integrated Vision – Caspeele, Taerwe & Frangopol (eds), Taylor & Francis Group, London, pp. 1769–1778

Stroup DW, Bryner NP (2007) Structural Collapse Research at NIST, 13 pages

Thoma K (2004) Stochastische Betrachtung von Modellen für vorgespannte Zugelemente, IBK Bericht Nr. 287, August 2004, Institut für Baustatik und Konstruktion, ETH Zürich, Zürich

United Nations (2018) Tracking Progress Towards Inclusive, Safe, Resilient and Sustainable Cities and Human Settlements. DG 11 Synthesis Report 2018, Nairobi, Kenia

Usbeck T, Wohlgemuth T, Chr P, Volz R, Benistone M, Dobbertin M (2010) Wind speed measurements and forest damage in Canton Zurich (Central Europe) from 1891 to winter 2007. Int J Climatol 30:347–358

VKF (2018) Brandschutzexperte und Risikospezialist für das Projektteam „VKF-Brandschutzvorschriften 2026" (Erarbeitung der VKF-Brandschutzvorschriften 2026), Verein Kantonaler Feuerversicherungen, Bern

Vogel T, Zwicky D, Joray D, Diggelmann M, Hoj NP (2009) Tragsicherheit der bestehenden Kunstbauten, Sicherheit des Verkehrssystems Strasse und dessen Kunstbauten, Bundesamt für Strassen, Dezember 2009, Bern

Wallace WL (1971) The Logic of Science in Sociology. Aldine de Gruyter, Abingdon, New York

Wang Z (2008) A technical note on seismic microzonation in the central United States. J. Earth Syst Sci 117(S2):749–756

Wenk T (2005) Beurteilung der Erdbebensicherheit bestehender Strassenbrücken. Bundesamt für Strassen – ASTRA, Bern

Wetter O, Ch, Pfister, Weingartner R, Luterbacher J, Reist T, Trösch J (2011) The largest floods in the High Rhine basin since 1268 assessed from documentary and instrumental evidence. Hydrol Sci J 56(5):733–758

Wheatley S, Sovacool B, Sornette D (2015) Of Disasters and Dragon Kings: A Statistical Analysis of Nuclear Power Incidents & Accidents, arXiv:1504.02380v1. Accessed 7. April 2015, 24 pages

Yanev PI, Yanev AJ (2013) Earthquake Engineering Lessons for Nuclear Power Plant Structures, Systems, and Components from Fukushima and Onagawa. 8th Nu-clear Plants current Issues Symposium, 23–25.1.2013, Orland, Florida, US

Zanini MA, Faleschini F, Pellegrino C (2017) Bridge life-cycle prediction through visual inspection data updating, Life Cycle of Engineering Systems: Emphasis on Sustainable Infrastructure – Bakker, Frangopol & van Breugel (eds), Taylor & Franics, London, pp 1518–1525

ZIA (2017) Zentraler Immobilien Ausschuss e.V., Immobilienwirtschaft 2017, Berlin

Preliminary Considerations

2

2.1 Introduction

The subjects of this book are buildings and structures, collapses, collapse frequencies and failure probabilities. The four terms are briefly explained in this chapter.

In addition, the book also uses the definitions of the structure types of bridges, dams, tunnels, retaining structures, buildings, stadiums, and wind turbines. The definitions of these terms can be found in Chaps. 3 to 9. In addition, the necessary terms for nuclear power plants, which are used for comparison, are discussed in Chap. 10.

2.2 Definition of Structures

There are different variants of the definition of a building and a structure respectively. An example of a definition is provided by the German Civil Code (BGB 2002): A building and structure respectively is *"... a thing that is firmly connected to the ground and has been produced by material and labour (by people). A building [and structure respectively] is an integral part of the property and can only be separated from the property again with damage or a change in its nature"*.

The German building regulations know the term "structural installations". The Saxon Building Code (SächsBO 2004) defines this term: *"Structural installations are installations connected to the ground and made of building products"*.

Wikipedia (Surhone et al. 2011) states *"A structure is a man-made construction in dormant contact with the ground. It is usually designed for long-term use."*

In addition, very specific definitions exist, such as according to SIA 465 (1998) "*Structures are technical systems/products with a uniform safety management*" or very

D. Proske, *The Collapse Frequency of Structures*,
https://doi.org/10.1007/978-3-030-97247-9_2

general ones, such as in Eurocode 0 (2017) *"... everything that is created by construction or results from construction work"*.

Structures show the following characteristics:

- They are built by man.
- They usually have a longer scheduled service life.
- They consider of structural elements which are either space-defining (wall, ceiling), space-spanning (bridge, tower) or form-giving (monuments).
- They have force-conducting elements, such as walls, ceilings, cables, beams.
- They are immobile. Thus, a structure can only be removed from a property by changing its nature.
- Structures are usually unique.

In principle, structures can be divided into two classes (Berner et al. 2013):

- Buildings as residential, commercial, and public buildings as well as structures for the sports and leisure sector such as stadiums and arenas respectively and
- Civil engineering structures as part of the infrastructure with bridges, dams, retaining walls and tunnels.

The Brockhaus Encyclopaedia (2017) defines civil engineering structures as *"structures that require the solution of technical-constructive and static tasks as well as soil-mechanical problems in order to be erected."*

Wikipedia (2015) defines civil engineering structures as *"...Structures such as bridges, tunnels, dams, reservoirs, viaducts, etc., which are primarily designed and constructed by civil engineers, as special safety reserves must be available here. This requires extensive static calculations in advance."*

Both DIN 1076 (1999) and HOAI (2020) recognise the concept of civil engineering structures. Section 3 of the HOAI (2020) contains a list of structures that are considered engineering structures. DIN 1076 (1999) also contains a list and a definition for additional engineering structures.

In general, civil engineering structures for traffic routes can be divided into terrain fills (embankments), terrain cuts (retaining structures), terrain overpasses (bridges) and terrain cuts (tunnels) to meet the required gradients. This is shown symbolically in Table 2.1. These four types of structures are dealt with in the context of this book, whereby dams are considered here as retaining structures, not as part of the roadway.

2.3 Definition of Collapse

The definition of the collapse of a structure is unfortunately not uniform. Some examples are given below.

Table 2.1 Typological classification of roadway engineering structures (Hofmann et al. 2021)

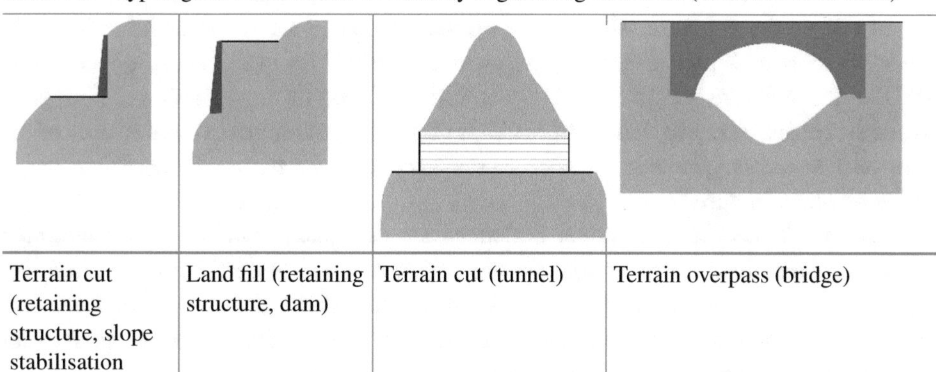

Terrain cut (retaining structure, slope stabilisation	Land fill (retaining structure, dam)	Terrain cut (tunnel)	Terrain overpass (bridge)

Wardhana and Hadipriono (2003) define a bridge collapse as follows: *"A bridge collapse occurs when an entire bridge or a substantial part comes down, at which point the structure loses its ability to perform its function."*

Imhof (2004) defines collapse as *"… render the structure incapable of remaining in service."*

OSHA (2021) defines collapse as: *"When internal load bearing structural elements fail, a building will collapse into itself and exterior walls are pulled into the falling structure. This scenario may be caused by construction activity, an earthquake, or fire, and may result in a dense debris field with a small footprint. Alternatively, if the structural failure is caused by an explosion or natural forces such as weather, the building may collapse in an outward direction, resulting in a less dense and more scattered debris field."*

The NYSDOT (2004; taken from Cook 2014) has introduced two definitions: total collapse and partial collapse. The definition of a complete collapse is *"structures which all primary members of a span or several spans have undergone severe deformation such that no travel lanes are passable."* The definition of a partial collapse is *"structures on which all or some of the primary structural members of a span or multiple spans have undergone severe deformation such that the lives of those travelling on or under the structure would be in danger."*

Other authors see a collapse as the development of a kinematic chain (Adam et al. 2018; Leyendecker and Ellingwood 1997). Exceeding an ultimate limit state as in Eurocode 0 (2017) or SIA 260 (2013) alone is not sufficient for this because the structure can remain in a state of possible repair (Starossek 2005; Klingmüller and Bourgund 1992). In the field of seismic, there are further limit states (FEMA 1997; Gkatzogias and Kappos 2015, see Fig. 1.3). For example, there is an explicit limit state of repairable damage. At this limit state, there is still no danger to life and limb. In the case of earthquakes, the extent of damage can be defined by the loss of volume of a building due to collapse. (Coburn et al. 1992; taken from Hingorani et al. 2020)

Consequently, a collapse is a sequence of failure of structural elements leading to the destruction of the structure. The author considers a collapse to be a condition in which the structure and its parts can no longer maintain themselves in their position and form, and thus the structure is destroyed. This definition is based on the definition of a supporting structure according to SIA 260 (2013). It states that a structure is *"the totality of the components and the foundation soil that are necessary for the equilibrium and shape retention of a structure."*

Dam failure, on the other hand, is defined as: *"the sudden rapid and uncontrolled release of impounded water or liquid-borne solids. It is recognized that there are lesser degrees of failure and that any malfunction or abnormality outside the design assumptions and parameters that adversely affect a dam's primary function of impounding water could be considered a failure."* (FEMA 2015)

2.4 Calculation of Collapse Frequencies

The collapse frequency of structures F results from the ratio of the number of collapsed structures n_c in relation to the population of all structures n:

$$F = \frac{n_c}{n} \tag{2.1}$$

The three parameters refer to a reference area, a specific reference period and a observation time respectively. As a rule, one year is often used as the reference period for the number of collapses n_c. The reference time for determining the total number of structures n depends on the respective operator or author. The reference area often refers to national borders, but it can also be administrative borders of infrastructure operators. For certain actions, e.g. earthquakes, an adjustment of the number of collapsed structures to the exposed area is usually made.

This simple formula does not take time-dependent statistical parameters into account. The temporal independence, the ergodicity, is not explicitly tested, but trend investigations are carried out. Furthermore, the observation period depends on the available data.

2.5 Calculation of Failure Probabilities

Probabilistic methods are now an integral part of modern building regulations. Overall, the first development steps of this procedure are almost 100 years old (Mayer 1926). The methods have been increasingly used since the 1970s.

The aim of probabilistic calculations is to determine the probability of failure P_f or an equivalent substitute measure (safety index β). Input variables of such calculations are random variables that obey probability density functions. Through arbitrary functional correlations, the composite probability density of these random variables is split into two parts. This functional relationship is called the limit state equation. It includes, for example, a load bearing or serviceability structural analysis without safety elements.

The probability of failure P_{f} is calculated as follows:

$$P_f = \iint\limits_{g(\boldsymbol{X})<0} f_{\boldsymbol{X}}(\boldsymbol{x})d\boldsymbol{x} \tag{2.2}$$

where $g(\boldsymbol{X})$ is the limit state equation, $\boldsymbol{X}$ is the vector of design variables and $f_{\boldsymbol{X}}(\boldsymbol{x})$ is the composite probability density of the vector $\boldsymbol{X}$. The limit state equation $g(\boldsymbol{X})$ represents the integration limit over the space of design variables.

Figure 2.1 visualises this relationship, also showing the development of the probability functions from measurements of actions, e.g. variable actions (steps Ⓐ and Ⓑ), and from measurements of resistance values, e.g. strengths (steps Ⓒ and Ⓓ).

The safety index β is often used instead of the probability of failure. This is directly linked to the probability of failure using the standard normal distribution Φ:

$$P_f = \Phi(-\beta) \tag{2.3}$$

Today, various commercial and non-commercial programmes are available for performing such calculations, which in turn use different calculation methods, such as FORM (First Order Reliability Method), SORM (Second Order Reliability Method) or Monte Carlo simulation and its variants (Spaethe 1992; Melchers 1999). The probability of

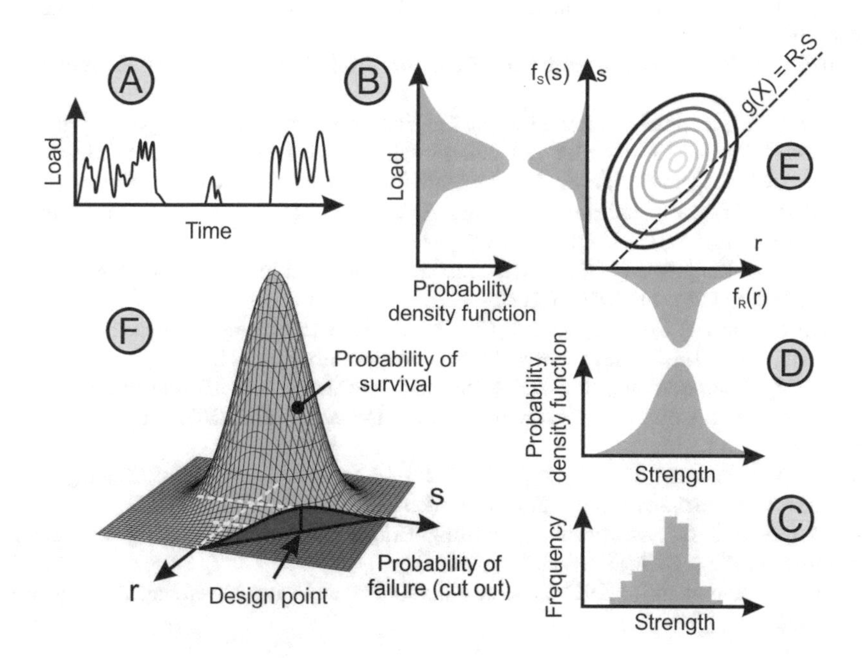

Fig. 2.1 Simplified representation of the determination of the probability of failure (Proske 2009)

failure is referred to in the literature (Spaethe 1992; Melchers 1999) as the operational probability of failure in order to clearly indicate the difference between calculation (probability of failure) and observation (collapse frequency and collapse probability respectively).

Here, too, a year is usually used as the reference period. In contrast to the collapse frequency, the failure probability in most cases only refers to a load-bearing capacity structural analysis at a specific position, not to the overall load-bearing capacity of the structure. This must then be done by linking individual failure probabilities according to certain rules such as Rackwitz- and Ditlevsen-limits respectively.

References

Adam JM, Parisi F, Sagaseta J, Lu X (2018) Research and practice on progressive collapse and robustness of building structures in the 21st century. Eng Struct 173(15):122–149

Berner F, Kochendürfer B, Schach R (2013) Grundlagen der Baubetriebslehre. Band 1: Baubetriebswirtschaft, Leitfaden des Baubetriebs und der Bauwirtschaft B.G. Teubner Verlag/ GWV Fachverlage GmbH, Wiesbaden

BGB (2002) Bürgerliches Gesetzbuch, 2. Januar 2002

Brockhaus (2017) Ingenieurbauwerke. https://brockhaus.de

Coburn A, Spence R, Pomonis A (1992) Factors determining human casualty levels in earthquakes: mortality prediction in building collapse. Earthquake Engineering, Tenth World Conference. Madrid

Cook W (2014) Bridge Failure Rates, Consequences, and Predictive Trends. Dissertation, Utah State University Logan

DIN 1076 (1999) Ingenieurbauwerke im Zuge von Straßen und Wegen – Überwachung und Prüfung

Eurocode 0 (2017) Basis of Structural Design, Englischsprachige Fassung, Ausgabe April 2017

FEMA (1997) FEMA 273 – NEHRP Guidelines for the seismic rehabilitation of buildings, Federal Emergency Management Agency, October 1997, Washington D.C., Redwood City

FEMA (2015) Federal Guidelines for Dam Safety Risk Management, Federal Emergency Management Agency, Washington DC, 332 pages

Gkatzogias KI, Kappos AJ (2015) Deformation-based seismic design of concrete bridges. Earthquakes and Struct 9(5):1045–1067

Hingorani R, Tanner P, Prieto M, Lara C (2020) Consequence classes and associated models for predicting loss of life in collapse of building structures. Struct Saf 85:1–13

HOAI (2020) Honorarordnung für Architekten und Ingenieure vom 10. Juli 2013 (BGBl. I S. 2276), die durch Artikel 1 der Verordnung vom 2. Dezember 2020 (BGBl. I S. 2636) geändert worden ist

Hofmann C, Proske D, Zeck K (2021) Vergleich der Einsturzhäufigkeit und Versagenswahrscheinlichkeit von Stützbauwerken. Bautechnik 98(7):475–481.

Imhof D (2004) Risk Assessment of Existing Bridge Structures, University of Cambridge, Dissertation, Kings College, December 2004

Klingmüller O, Bourgund U (1992) Sicherheit und Risiko im Konstruktiven Ingenieurbau. Vieweg, Wiesbaden

Leyendecker EV, Ellingwood BR (1977) Design Methods for Reducing the Risk of Progressive Collapse in Buildings, U.S. Department of Commerce, National Bureau of Standards Buildings Science Series 98, Washington, D.C., April 1977

Mayer M (1926) Die Sicherheit der Bauwerke und ihre Berechnung nach Grenzkräften anstatt nach zulässigen Spannungen. Verlag von Julius Springer, Berlin

Melchers RE (1999) Structural reliability analysis and prediction. Wiley

NYSDOT 2004 Bridge Safety Assurance: Hydraulic Vulnerability Manual. New York State Department of Transportation

OSHA (2021) Structural Collapse Guide, Occupational Safety and Health Administration, United States Department of Labor. https://www.osha.gov/SLTC/emergencypreparedness/guides/structural.html

Proske D (2009) Catalogue of risks. Springer, Heidelberg

SächsBO 2004) Sächsische Bauordnung, 28. Mai 2004

SIA 260 (2013) Grundlagen der Projektierung von Tragwerken, Schweizer Ingenieur- und Architektenverein, Zürich

SIA 465 (1998) Sicherheit von Bauten und Anlagen, Schweizer Ingenieur- und Architektenverein, Zürich

Spaethe G (1992) Die Sicherheit tragender Baukonstruktionen, zweite, neubearbeite. Springer, Wien

Starossek U (2005) Progressiver Kollaps von Bauwerken. Beton- und Stahlbetonbau 100(4):305–317

Surhone LM, Tennoe MT, Henssonow SF (2011) Bauwerk, Betaskript Publishing

Wardhana K, Hadipriono FC (2003) Analysis of Recent Bridge Failures in the United States, Journal of Performance of Constructed Facilities, ASCE, August 2003, pp 144–150

Bridges

3

3.1 Introduction

Bridge structures are an essential part of the infrastructure in almost all countries of the world. They ensure the functionality and thus the maintenance of modern human societies, which are characterised by high flows of information, energy and materials.

The volume of freight traffic on German roads has grown exponentially in weight and volume over the past decades (Naumann 2002). For example, Hannawald et al. (2003) observed a 70-tonne vehicle on a German motorway bridge during only one day of traffic measurement. Pircher et al. (2009) and Enright et al. (2011) report 100-tonne vehicles on roads. Further expansion of heavy goods vehicles is planned by the EU (European Commission Directive 96/53/EC). These developments directly affect the requirements for today's bridges.

Wooden bridges have been built for several thousand years and massive stone bridges for more than 2,000 years (Heinrich 1983). An indication of an early wooden bridge to cross Lake Zurich between Rapperswil and Hurden are pile remains with an age of 3,500 years.

The global stock of bridges grew significantly with the onset of industrialisation and the introduction of railways (Proske 2019c, see Fig. 3.1), during the period of growth in private transport in the 1960s and 1970s (Proske 2019c, see Fig. 3.2), and for the last two decades due to strong economic growth in Asia (Proske 2018b, see China in Fig. 3.3, for Japan see Fujino 2018).

D. Proske, *The Collapse Frequency of Structures*,
https://doi.org/10.1007/978-3-030-97247-9_3

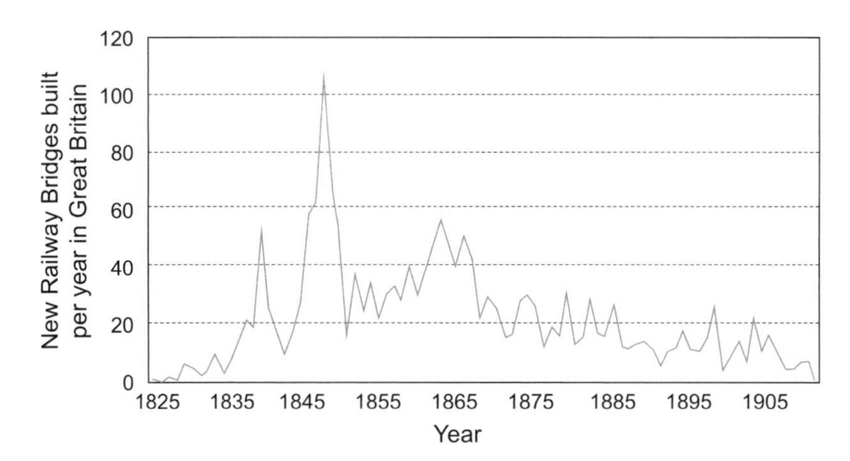

Fig. 3.1 Newly built railway bridges in Great Britain in the nineteenth century (Proske 2019c)

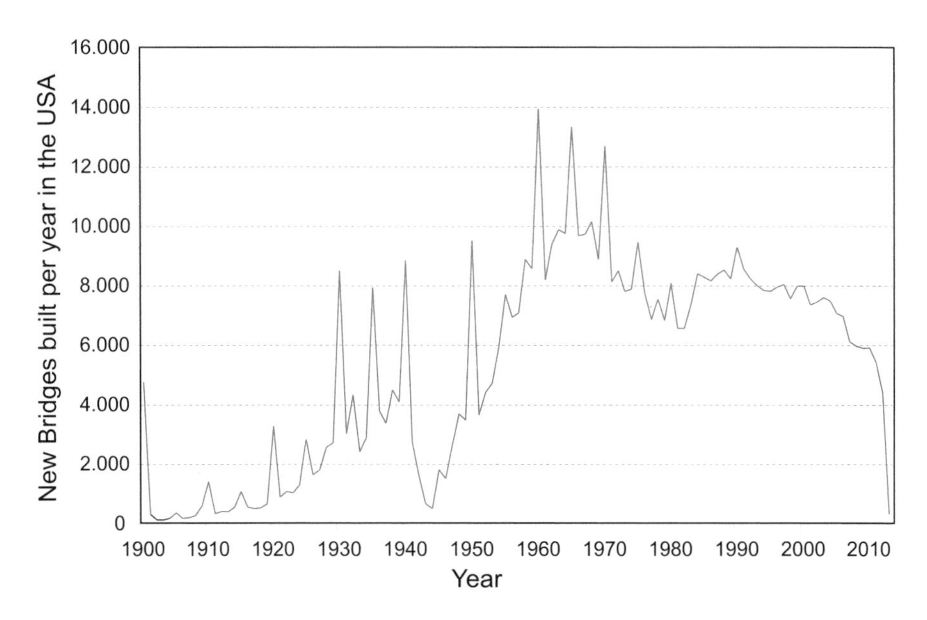

Fig. 3.2 Newly built bridges in the USA in the twentieth century (Proske 2019c)

3.2 Definition of Bridges

According to the Association of American State Highway and Transportation Officials (AASHTO), bridges are defined as *"a structure, including supports, erected over a depression or an obstruction (such as water, highway and railway) having a track or passageway for carrying traffic or other moving loads."* The definition was taken from Hersi (2009) and Imhof (2004).

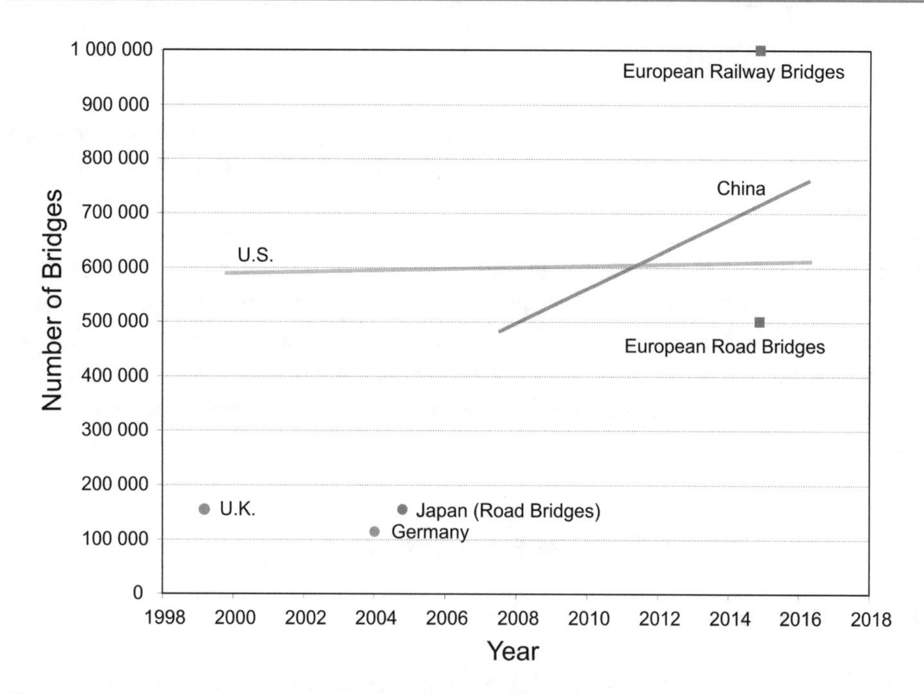

Fig. 3.3 Bridge stock in different countries (Proske 2017a, 2018a)

A definition in South Africa (Wolhuter 2015) reads: *"A bridge is a structure erected with a deck for carrying traffic over or under an obstruction and with a clear span of 6 m or more. Where the clear span is less than 6 m, reference is to a culvert."*

The National Bridge Inspection Standard gives the following definition: *"A structure including supports erected over a depression or an obstruction, such as water, highway, or railway, and having a track or passageway for carrying traffic or other moving loads, and having an opening measured along the centre of the roadway of more than 20 ft (6.1 m) between undercopings of abutments or spring lines of arches, or extreme ends of openings for multiple boxes; it may also include multiple pipes, where the clear distance between openings is less than half of the smaller contiguous opening."* (FHWA 1996).

According to the German DIN 1076 (1999), bridges are *"overpasses of a traffic route over another traffic route, over a body of water or lower-lying terrain, if their clear width measured at right angles between the abutments is 2.00 m or more"*.

The minimum span in particular therefore differs in different countries. This makes it difficult to compare the bridge stock and thus to determine the collapse frequency.

3.3 Stock of Bridges

The great need for transport has led to the fact that today there is one bridge for every 500 inhabitants in industrialised countries and for every 2,000 inhabitants in developing countries. With a world population of 7.6 billion people, the total number of all man-made bridges worldwide is probably between five and six million. Table 3.1 lists the inventory of bridges in different countries.

Of course, topography would be more appropriate for estimating the total number of all bridges. Weber (1999), for example, provides bridge densities (number of bridges per 10 km of track) related to railway lines for different countries with specific topography (Table 3.2). However, such data are not available to the same extent as the absolute figures in Table 3.1.

3.4 Calculation of Collapse Frequencies

3.4.1 Introduction

There are numerous databases on bridge collapses (Proske 2018b). Some of these include more than 1,000 collapses. Table 3.3 presents an extended compilation of such studies. Overall, it must be stated that there is an extraordinarily good data basis for bridges. Other publications dealing with bridge collapses that are not mentioned in Table 3.3 are Tweed (1969), Brown (1979), Akesson (2008), Diehm and Hall (2013), FHWA (2015) and Abels & Annes (2017).

However, many bridge collapses affect smaller bridges in the municipal area (Spector and Gifford 1986). These are rarely reported on a supra-regional basis, so there is a risk of systematic "underreporting" for the databases (Knoll 1965).

The evaluation of practically all databases shows that the main causes of bridge failure are floods and impacts (Proske 2018b). In this context, hundreds of bridges can be destroyed during an event, as in the 2011 tsunami in Japan (Proske 2020b), or over 4,000 bridges can be damaged, as in an earthquake in China in 2008 (Wang and Lee 2009). Table 3.4 lists various such extreme events.

Collapses under traffic, such as the Morandi Bridge in Italy (Genoa) in 2018, Taiwan (Nanfang'ao) in 2019 and France (Mirepoix-sur-Tarn) in 2019, rarely occur in industrialised countries.

3.4.2 Central Estimator

Figure 3.4 shows the mean collapse frequencies according to different publications and for different time periods. The averaging involves certain assumptions. These are

Table 3.1 Number of bridges in countries (based on Proske 2017a, 2018b, supplemented)

Country	Population in millions	Number of bridges	Reference
Australia	25	54,000	Caprani (2018) **
Brazil	210	4,725	Oliveira et al. (2019) ****
USA	260	600,000	Dunker (1993)
USA	319	607,380	ASCE (2013), FHWA (2017)
China	1,325	500,000	Yan and Shao (2008) **
China	1,344	689,400	Zhang et al. (2014)
China	1,371	750,000	Xu et al. (2016)
China	1,440	878,300	Statista (2020a) **
Germany	83	120,000	Der Prüfingenieur (2004)
Germany	83	150,000	Kroker (2013)
Europe	746	500,000	Casas (2015) **
Europe	746	1,000,000	Casas (2015) ***
Great Britain	58	100,000	Menzies (1996)
Great Britain	59	150,000	Woodward et al. (1999)
Italy		60,000	Weber (1999), UIC (2005), Proske and van Gelder (2009) ***
Italy		17,000	Borzi et al. (2015) **
Italy		1,000,000	Macbeth (2018)
Japan	127	155,000	MLIT (2005) *
Japan	128	600,000	Gordenker (2011)
Canada	38	47,000	Statistics Canada (2018) **
Korea	52	33,000	Chang and Choo (2009)
Korea	52	36,000	Statista (2020b) **
Russia	144	85,000	Hingorani (2021), Syrkov (2020)
Switzerland	8,4	18,000	Proske (2020)

* Only road bridges longer than 15 m
** Road bridges only
*** Railway bridges only
**** Motorway bridges only

explained in Proske (2018b). Trend lines and results of probabilistic calculations are also integrated into the figure.

The data for the determination of collapse frequencies in Imhof (2004) and McLinn (2010) are likely to be incomplete. This probably also applies to the results of Vogel et al. (2009) worldwide, since this study, on the one hand, uses results of Imhof and since, on the other hand, it does not seem comprehensible that the worldwide bridge

Table 3.2 Number of railway bridges according to Weber (1999)

Railway companies	Operating length in km	Number of railway bridges	Bridge density
Belgian (SNCB/NMBS)	3,432	3,400	10
British (BR)	16,528	26,240	16
Bulgarian (BDŽ)	4,299	982	2
Danish (DSB)	2,344	1,500	6
German (DB)	4087	32,017	8
Finnish (VR)	5,874	1,905	3
French (SNCF)	32,731	28,259	9
Greek (CH, OSE)	2,484	21,000	8
Italian (FS)	16,112	59,473	37
Irish (CIE)	1,944	2,752	14
Yugoslav (JŽ)		2,770	
Luxembourgish (CFL)	275	282	10
Dutch (NL)	2,753	2,790	10
Norwegian (NSB)	4,027	2,700	7
Austrian (ÖBB)	5,605	5,048	9
Polish (PKP)	25,254	8,500	3
Portuguese (CP)	3,054	1,928	6
Rhaetian Railway (RhB)	375	489	13
Romanian (CFR)	11,430	4,067	4
Swedish (SJ)	9,846	3,500	4
Swiss (SBB)	2,985	5,267	18
Spanish (RENFE)	13,041	6,371	5
Czechoslovak (ČSD)	13,100	9,411	7
Hungarian (MAV)	7,605	2,375	3

collapse frequency is almost a power of ten better than the Swiss bridge collapse frequency. Very credible are the data of the USA bridge collapses, which lie in a narrow corridor. Overall, the observed collapse frequencies are just below or just above 10^{-4} per year.

3.4.3 Measure of Deviation

In addition to determining mean collapse frequencies, one can also determine the dispersion of the data, both for aleatory and epistemic uncertainty. The latter is shown in Table 3.5 as the confidence interval of the mean, the former as the standard deviation. The

Table 3.3 List of studies investigating bridge collapses (mostly from Proske 2018b, updated)

Reference	Region	Period	Number of bridge collapses
Breysse and Ndiaye (2014)	World	n/a	n/a
Bridge Forum (2017)	World	1444–2009	360
Brückenweb (2017): Accidents and collapses	World	1157–2017	284
Christian (2010)/Briaud et al. (2012)	USA	1966–2005	1,502
Cook (2014)	USA	1987–2011	103
Deng et al. (2016)	China		
Diaz et al. (2009)	Colombia	1986–2008	63
Dubbudu (2016): only fatalities given (297)	India	2010–2014	n/a
Fard (2012)		1818–2012	100
Fu et al. (2012)	China	2000–2012	157
Garg et al. (2022)	India	1977–2017	2,010
Harik et al. (1990)	USA	1951–1988	77
Hersi (2009)	USA	2000–2008	161
Imam and Chryssanthopoulos (2012): Steel bridges	World	early 19th c. -2011	164
Imhof (2004)	World	1444–2004	347
Lee et al. (2013a)	World	1980–2012	1,062
Lee et al. (2013b)	World	1876–2005	1,723
McLinn (2010)	USA	before 1900	40 per year
McLinn (2010): Only "*big*" collapses	World	1970–2009	71
Menzies (1996)	Great Britain	n/a	n/a
Montalvo and Cook (2017)	USA	1992–2014	428
Schaap and Caner (2021)	Turkey	2000–2019	80
Scheer (2000)	World	813–2008	536
Sharma (2010)	World	1800–2009	1,814
Sharma and Mohan (2011)	USA	1800–2009	1,367
Smith (1976)	USA	1847–1975	143
Taricska (2014)	USA	2000–2012	341
Vogel et al. (2009)	Switzerland	with data from Imhof	
Vogel et al. (2009)	World	with data from Imhof	
Wardhana and Hadipriono (2003)	USA	1989–2000	503

(continued)

Table 3.3 (continued)

Reference	Region	Period	Number of bridge collapses
Wikipedia (2017)	World	1297–2017	242
Xu et al (2016)	China	2000–2014	302
Zerna (1983): Steel bridges	USA	before 1900	1
Zerna (1983): Suspension bridges	World	1900–1940	7
Syrkov et al., Hingorani et al. (2022), IABSE TG1.5	World	1965–2020	805

Table 3.4 Examples of short-term large-scale events with a significant number of bridge collapses or damages (Proske 2020b)

Year	Location	Description or number in the literature	Initiating Event
1784	Central Europe	22 bridges destroyed or heavily damage	Flood
1947	USA	"Great losses" on bridges	Flood
1952	GB	28 bridges damaged and destroyed	Flood
1964/1972	USA	383 bridges destroyed or damaged	Flood
1976	Japan	233 bridges destroyed or damaged	Typhoon and flooding
1985	USA	73 bridges destroyed	Flood
1987	USA	17 bridges destroyed	Flood
1993	USA	110 bridges destroyed	Flood
1998	Bangladesh	400 bridges damaged	Flood
2002	Saxony, Germany	>450 bridges damaged, >15 bridges destroyed	Flood
2005	USA	70 bridges destroyed	Storm Katrina and floods
2008	China	"large number" of bridges collapsed (landslides), 4,840 bridges damaged	Earthquake
2009	GB	7 bridges destroyed	Flood
2011	USA	40 bridges destroyed	Flood
2011	Japan	>300 bridges destroyed	Earthquake and tsunami
2012	Afghanistan	400 bridges destroyed	Flood
2015	GB	131 bridges damaged	Flood
2021	Germany	52 bridges destroyed or partly destroyed	Flood

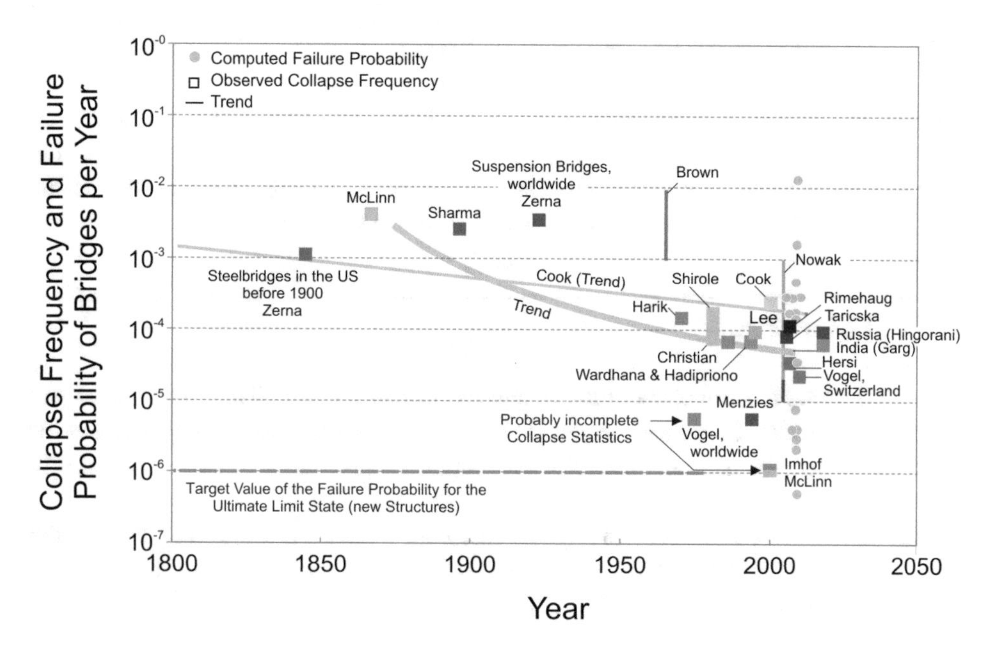

Fig. 3.4 Collapse frequencies and failure probabilities of bridges (Proske 2020a)

aleatory uncertainty is about one power of ten, both for the observed collapse frequency and for the calculated failure probability. This extraordinarily high value is at least partly due to the long observation period for the collapse frequencies.

3.4.4 Trend

Linear and exponentially decreasing trend functions of the mean observed collapse frequencies can be found in the literature (Proske 2018b). The decadal logarithm of the mean change per century is about 0.67. This means that the collapse frequency has decreased by a factor of 4.67 per century.

Table 3.5 Mean values including confidence intervals and standard deviations (Proske 2020a)

Probabilistic calculations		Observed frequency		Target value per year
Standard deviation	Mean values incl. confidence interval	Standard deviation	Mean values incl. incl. confidence interval	
10.93	2.00×10^{-4}, (1.92×10^{-4}, 2.07×10^{-4})	9.15	1.17×10^{-4}, (1.13×10^{-4}, 1.22×10^{-4})	10^{-6}

3.5 Calculation of Failure Probabilities

To compare the observed collapse frequencies and the calculated failure probabilities, 16 probabilistic calculations were selected. These were selected as representative of the bridge stock in terms of building materials, bridge type and age. However, due to the small number of calculations used, a complete adjustment is not possible.

3.6 Comparison

Figure 3.5 shows a comparison of the frequency distribution of collapse frequencies and failure probabilities. Of course, the sample sizes differ considerably. While databases for collapses sometimes include over 1,000 samples, only just under 20 failure probability values were used.

Nevertheless, it can be seen in Fig. 3.5 that both parameters agree well both in the extreme values and in the shape of the distribution. As Table 3.5 shows, the standard deviations also coincide.

3.7 Causes of Collapse

Causes of collapses can be assigned to different thematic groups. Thus, for all building collapses, and this also applies to bridge collapses, human errors in planning, construction and use can be named as causes.

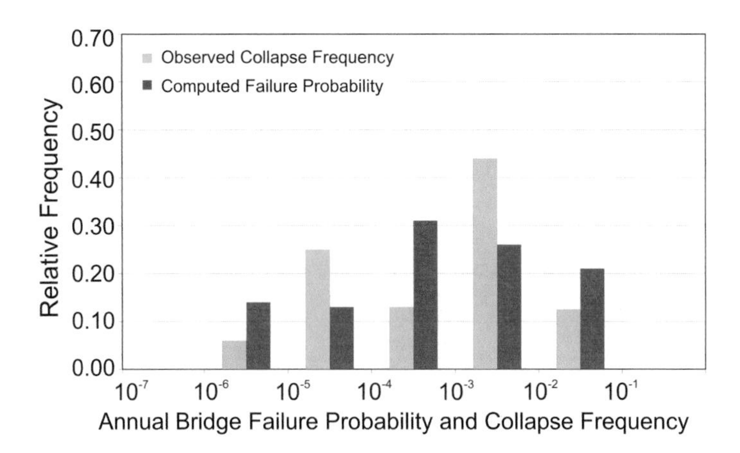

Fig. 3.5 Histogram of the calculated failure probabilities and the observed collapse frequencies for bridges (Proske 2020a)

In the same way, however, the causes of collapse are also associated with certain actions. Figure 3.6 shows such an allocation of the causes of collapse to different actions and by different authors. Practically all authors see flooding with the accompanying phenomenon of scour as the main cause of bridge collapses. This finding is found throughout the literature, other literature examples include Melville and Sutherland (1988), Gee (2003) and Lee et al. (2019). Table 3.4 also indirectly confirms this. Table 3.6 shows the proportions of bridge collapse causes related to the different bridge types (see also Biezma and Schanack 2007). Here, too, the cause of collapse is dominated by flooding in beam and arch bridges, which accounts for the majority of all bridges.

Figure 3.7 shows the causes of collapse as a function of bridge age as a box-whisker diagram. It can be seen that impacts tend to occur more in the young bridge age and overloading more in the older bridge age.

3.8 Mortality and Casualty Figures

The actual goal of the safety concept in construction is the protection of human life and indirectly the protection of assets. The collapse frequency describes a loss of assets as damage. However, the recording of direct and indirect damages is very time-consuming, as Fig. 3.8 shows. Therefore, here only the numbers of victims of and mortalities caused by bridge collapses will be discussed.

The first thing to note is that the percentage of bridge collapses resulting in fatalities is in the low single digit range, probably around 3%. In other words, the majority

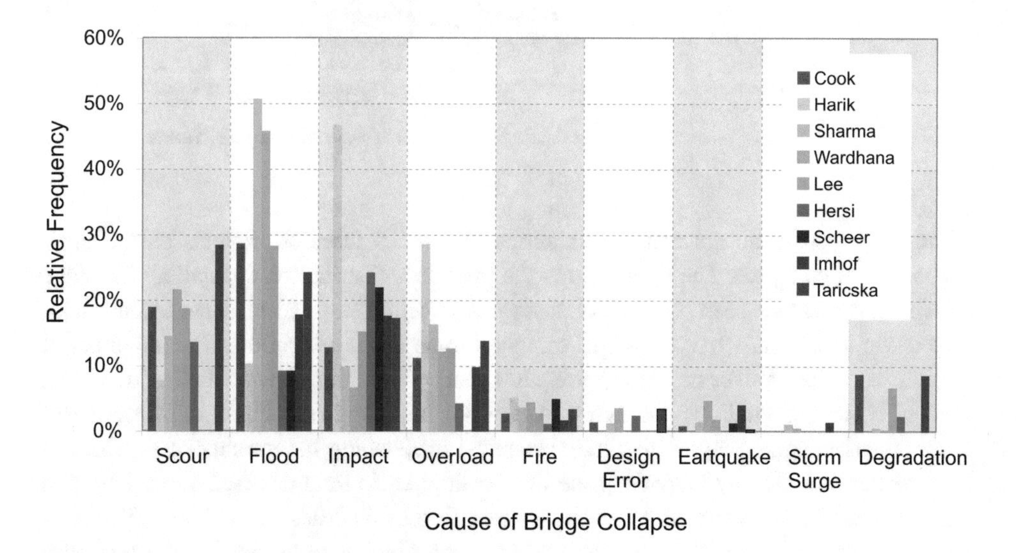

Fig. 3.6 Causes of bridge collapses according to various references (Proske 2018a, b)

Table 3.6 Causes of bridge collapses according to Proske (2020b) based on Deng et al. (2016)

Bridge type	Main causes of collapse	Share
Beam bridges	Floods and silting up	0.6
	Collision, earthquake	0.3
	Overloading, lack of maintenance	0.1
Stone arch bridges	Floods and silting up	0.6
	Collision, earthquake	0.3
	Overloading, lack of maintenance	0.1
Steel arch bridges	Overloading, lack of maintenance	0.5
	Storms	0.5
Steel truss bridges	Overloading, lack of maintenance	0.5
	Fatigue	0.5
Suspension and cable-stayed bridges	Storms	0.5
	Overloading, lack of maintenance	0.5

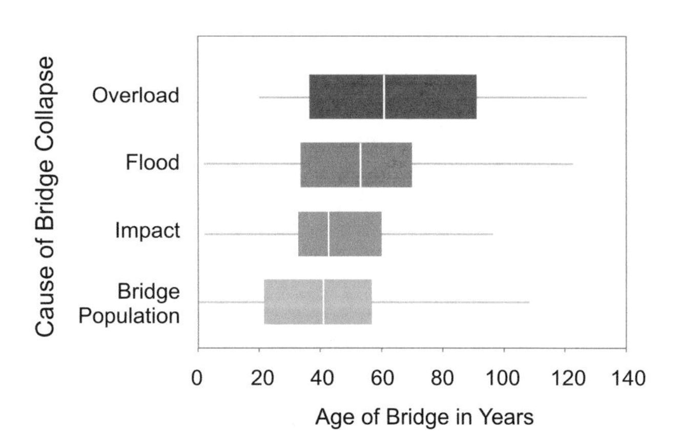

Fig. 3.7 Causes of bridge collapse depending on bridge age (Montalvo and Cook 2017)

of bridge collapses do not result in fatalities. Figure 3.9 therefore reflects only a small proportion of collapses. The figure shows the frequency distribution of fatalities in bridge collapses. As can be seen, bridge collapses can nevertheless lead to considerable numbers of victims. Figure 3.10 shows the temporal trend of the annual maximum number of victims per collapse over the last approx. 200 years. While this shows a falling trend, the ratio of injured to fatalities has unfortunately not improved (see Fig. 3.11). In contrast to earthquake engineering, which has succeeded in reducing the number of fatalities in earthquakes and thereby increasing the ratio of injured to fatalities (see Table 7.5), this has not been achieved for bridge collapses resulting in fatalities. If vehicles fall from a collapsing bridge or if vehicles under a bridge are hit by the collapsing bridge, mortality remains high.

I. Agency Consequences Related to the Element

1. Cost of special inspection of the element (Ce1),
2. Cost of the element demolition and removal of the debris (Ce2), and
3. Cost of the element replacement (Ce3).

II. Agency Consequences Related to the Bridge

4. Cost of special inspection (Cb1) to determine levels of damage of other components of the bridge as a result of the element failure,
5. Cost of maintenance and repair (Cb2) of other elements in the bridge that need repair as a result of the element failure,
6. Cost of demolition and replacement (Cb3) of other damaged elements in the bridge as a result of the element failure,
7. Cost of bridge strengthening (Cb4) that may be needed as a result of the element failure,
8. Cost (Cb5) of fixing or replacing utilities, lightings, and traffic signals on the bridge,
9. Cost of traffic management (Cb6) due to detouring that result from the element failure,
10. Consequences (Cb7) that may result in critical scour conditions due to the failure of some bridge elements,
11. Consequences (Cb8) related to load posting, and user-request permitting,
12. Consequences (Cb9) on the highway network due to congestion resulting from the element failure.

III. Consequences to Bridge Users

13. Cost of travel time delay (Cu1) of bridge users and other travelers in the highway network as a result of the element failure. Time delay could be due to traffic congestion, detours, or closures of bridge lanes. Travel time cost includes costs of time lost due to rerouted or diverted traffic travel, congestion that create queuing delays or stopping, or traffic delays that result from lane closures or work zones,
14. Cost of operating the vehicles (Cu2) of bridge users and other travelers in the highway network due to detours and traffic delays that result from the element failure. Vehicle operating cost includes increased gas or fuel consumption, maintenance, and depreciation of vehicles, and
15. Cost of damage to the vehicles (Cu3) of the bridge users as a result of the bridge element failure.

IV. Consequences Related to Traffic Accidents

16. Cost of traffic accidents due to vehicle collisions with the bridge (Ca1) as a result of the element failure,
17. Cost for load tests (Ca2) to determine the damage due to accidents that result from the element failure,
18. Cost of specific actions (Ca3) needed to repair or replace damaged components of the bridge due to accidents that result from the element failure,
19. Cost of damages to vehicles and other properties (Ca4) due to accidents resulting from the element failure,
20. Cost of removal of damaged cars and debris (Ca5) due to accidents resulting from the element failure,
21. Cost of emergency services and police officers (Ca6) needed for accidents resulting from the element failure,
22. Cost of insurance (Ca7) for accidents resulting from the element failure,
23. Cost of travel time delay (Ca8) due to the emergency services for accidents resulting from the element failure, and
24. Cost of congestion in the highway network (Ca9) due to accidents resulting from the element failure.

V. Consequences Relating to Health and Safety

25. Possible losses of human lives (Ch1),
26. Possible body injuries (Ch2),
27. Medical care expenses (Ch3),
28. Legal expenses (Ch4),
29. Insurance expenses (Ch5),
30. Lost of productivity of affected people (Ch6),
31. Cost of pain and suffering (Ch7),
32. Cost of loss of enjoyment of life (Ch8),
33. Loss of future earnings (Ch9), and
34. Cost of emergency services (Ch10).

VI. Consequences Relating to the Environment

35. Increased air pollution (Cenv1) due to fuel emissions of delayed or diverted vehicles in traffic congestion that result from the element failure (Hawk, 2003),
36. Impacts on water quality (Cenv2) in flowing streams or rivers under or adjacent to the bridge due to pollutants or waste products that result from the element failure (Hawk, 2003),
37. Disturbance to the agricultural land (Cenv3),
38. Impacts on the plants, trees, and forests (Cenv4),
39. Disposal of waste material and debris (Cenv5) that result from the element failure,
40. Noise and dust (Cenv6) due to the element failure,
41. Environmental damage (Cenv7) caused by spillage of hazardous material from vehicles on and under the bridge as a result of the element failure, and
42. Environmental damage (Cenv8) caused by fire and chemical spills resulting from traffic collisions with bridges as a result of the element failure.

VII. Consequences to Nearby Businesses

43. Loss of revenue (Cnb1) to the nearby businesses due to the element failure,
44. Loss of productivity (Cnb2) to the nearby businesses as a result of the element failure,
45. Cost of possible damages of surrounding properties (Cnb3) due to the element failure,
46. Cost of delay of services (Cnb4) to the nearby businesses due to the element failure, and
47. Cost of business travel (Cnb5) due to detours that result from the element failure.

VIII. Consequences to the General Public

48. Consequences on the public due to possible closure of the bridge (Cp1) as a result of the element failure,
49. Consequences on the public due to congestion in the highway network (Cp2) as a result of the element failure,
50. Damages to the society and general public due to public relation costs (Cp3),
51. Disturbance of emergency services (Cp4), and
52. Consequences on access to schools, libraries, health care facilities and governmental agencies (Cp5)

Fig. 3.8 Input parameters for determining the damage costs of bridges (Al-Wazeer 2007)

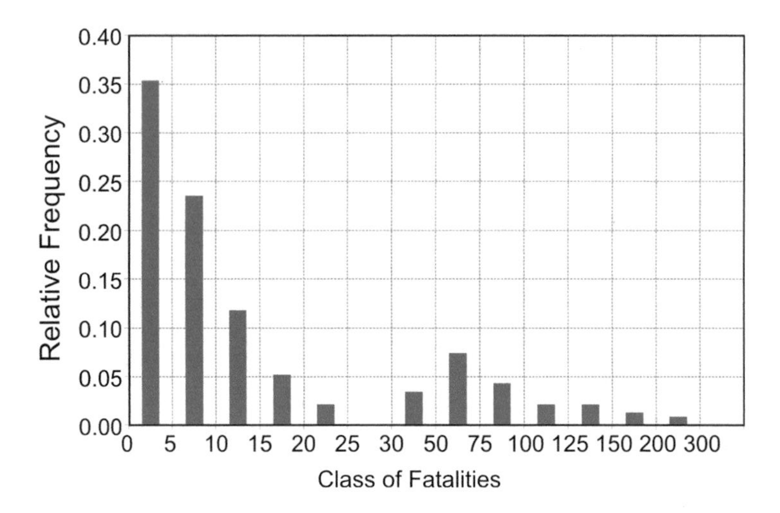

Fig. 3.9 Frequency distribution of bridge collapses with fatalities (Proske 2020b)

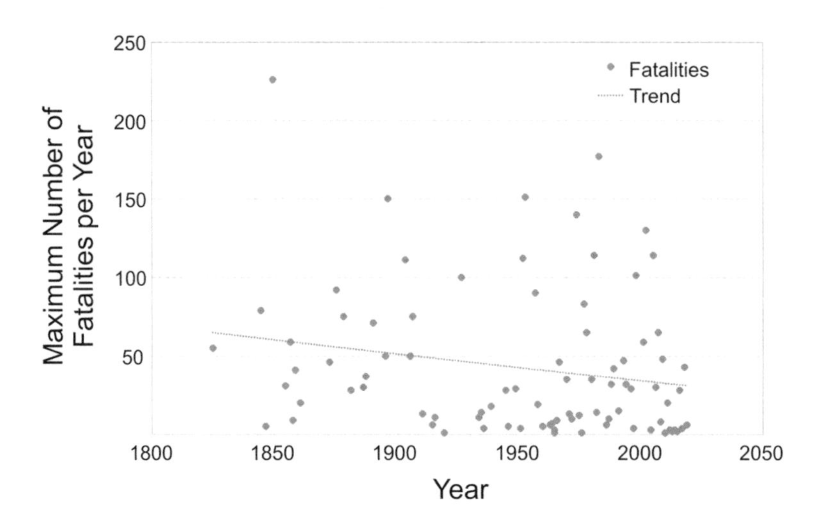

Fig. 3.10 Maximum number of fatalities due to bridge collapses per year (Proske 2020b)

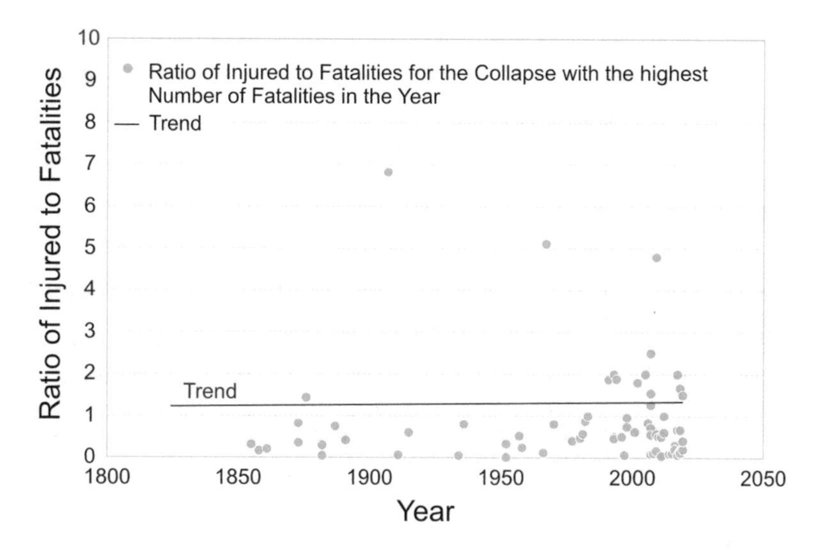

Fig. 3.11 Ratio of injured to fatalities in bridge collapses with fatalities (Proske 2020b)

Table 3.7 shows, however, that overall, the mortalities and other risk parameters for bridges are very low and well below the target values.

3.9 Summary

In summary

- The average collapse frequencies of bridges can be estimated robustly and reliably due to the extensive data available.
- The failure probabilities used for comparison purposes are representative but have a much smaller sample size.
- Nevertheless, both mean values and standard deviations of the observed collapse frequencies and the calculated failure probabilities agree surprisingly well.
- The causes of collapse in terms of actions are known.
- Only a small proportion of bridge collapses involve fatalities. However, collapses with more than 200 fatalities have been observed.

Table 3.7 Different risk values for bridges (Proske 2009, 2020d, 2022; Das 1997; Vogel et al. 2009; Blockley 1980; Menzies 1996)

Risk parameters	Bridges	Target or comparative value
Mortality per year	$2 \times 10^{-9} - 10^{-8}$	10^{-6}
Fatal Accident Rate	0.00002	0.2 to 2.0
Loss of years of life	Seconds and minutes	Car: 200 days

- The risk values of bridges are extraordinarily low and far below the target values.
- This chapter is based on results and data from the publications Proske (2017a, b, 2018a, b, 2019a, b, c, 2020a, b, c).

References

Abels & Annes PC (2017) Bridge Collapse Accidents, https://www.daveabels.com/chicago-construction-accident-lawyers/bridge-collapse/

Akesson B (2008) Understanding Bridge Collapses. CRC Press & Taylor and Francis, London

Al-Wazeer AAR (2007) Risk-based Bridge Maintenance Strategies, Dissertation, University of Maryland

ASCE (2013) 2013 Report card for America's infrastructure: Bridges: Overview, American Society of Civil Engineers. http://www.infrastructurereportcard.org

Biezma MV, Schanack F (2007) Collapse of steel bridges. J Perform Const Facil (ASCE) 21(5):398–405

Blockley DI (1980) The nature of structural design and safety. Wiley, Chichester

Borzi B Ceresa P Franchin P Noto F Calvi GM Pinto PE (2015) Seismic vulnerability of the Italian roadway bridge stock. Earthquake Spectra 31(4):2137– 2161

Breysse D & Ndiaye A (2014) Failure case databases related to risk in civil engineering. Proc Inst Civ Eng – Forensic Eng 167(1):27–37

Briaud JL, Gardoni P & Yao C (2012) Bridge scour risk. ICSE6 Conference Proceedings, Paris, August 27–31 2012, 1193–1210

Bridge Forum (2017) Bridge Failure Database, University of Cambridge. http://www.bridgeforum.org/dir/collapse/year/0000-3000.html

Brown CB (1979) A fuzzy safety measure, Journal of Engineering Mechanics Div. ASCE 105:855–872

Brückenweb (2017) Brückenkatastrophen – Einstürze – Unfall (Datenbank), F. Selke (Imprint). http://www.brueckenweb.de/2content/datenbank/katastrophen/3katastrophen.php

Caprani C (2018) Are Australian bridges safe, and can we do better? The Conversation, theconversation.com

Casas JR (2015) The bridges of the future or the future of bridges? Frontiers in Built Environment, pp 1–3. https://doi.org/10.3389/fbuil.2015.00003

Chang S-P, Choo J F (2009) Values of Bridge in the Formation of Cities. IABSE Workshop: Recent Major Bridges, Shanghai, China 11–12:25–46

Christian GA (2010) Bridge Failures – Lessons learned, Bridge Engineering Course, University at Buffalo, March 29 2010. http://mceer.buffalo.edu/education/bridge_speaker_series/2009-2010/presentations/P1%20Lessons%20learned%20from%20Bridge%20Failures_FINAL.pdf

Cook W (2014) Bridge Failure Rates. Diss, Utah State University Logan, Consequences and Predictive Trends

Das PC (1997) Safety of bridges. Thomas Telford, London

Deng L, Wang W, Yu Y (2016) State-of-the-art review on the causes and mechanisms of bridge collapse. J Perform Constr Facil 30(2):13

Der Prüfingenieur, News, April 2004, page 7

Diaz EEM, Moreno FN & Mohammadi J (2009) Investigation of Common Cause of Bridge Collapse in Colombia. Practice Periodical on Structural Design and Construction 14(4)

Diehm J, Hall K (2013) Bridge Collapses and structurally deficient bridges across the country. Huffington Post 24(5):2013

DIN 1076 (1999) Engineering structures in the course of roads and bridges, November 1999
Dubbudu R (2016) An average of 7 structures collapsed per day in the last 5 years, Factly: Making Public Data Meaningful, April 1 2016, https://factly.in/more-than-13000-lost-lives-instructure-collapses-in-the-last-5-years/
Dunker KF (1993) Why America's bridges are crumbling. Scientific American 266(3):18–25
Enright B, Caprani CC, O'Brien EJ (2011) Modelling of highway bridge traffic loading: some recent advances. In: 11th international conference on applications of statistics and probability in civil engineering (ICASP11), Zurich, 8 pages, on CD
Fard B (2012) A Comprehensive Study on 100 Bridge Failures and Their Reduction strategies. University of Buffalo, Civil Engineering
FHWA (1996) Recording and coding guide for the structure inventory and appraisal of the nation's bridges. Report FHWA-PD-96–001 Federal Highway Administration, US Department of Transportation, Washington DC
FHWA (2015) Engineering for Structural Stability in Bridge Construction, US DoT, Federal Highway Administration, National Highway Institute, NHI Course Number 130102, FHWA-HI-15–044, April 2015
FHWA (2017) National Bridge Inventory (NBI), https://www.fhwa.dot.gov/bridge/nbi.cfm
Fu Z, Ji B, Cheng M & Maeno H (2012) Statistical Analysis of the Causes of Bridge Collapse in China, Sixth Congress on Forensic Engineering
Fujino Y (2018) Bridge maintenance, renovation and managment - Research and Development of governmental program in Japan. In: Powers N, Frangopol DM, Al-Mahaidi R, Caprani C (Eds), Maintenance, Safety, Risk, Manamgent and Life-Cycle Performance of Bridges, Proceedings of the Ninth International Conference on Bridge Maintenance, Safety and Management (IABMAS 2018), Melbourne, Australia, CRC Press, Taylor and Francis, London, pages 2–14
Garg RK, Chandra S, Kumar A (2022) Analysis of bridge failures in India from 1977 to 2017. Struct Eng Int 18(3):295–312, https://doi.org/10.1080/15732479.2020.1832539
Gee KW (2003) Action: compliance with the national bridge inspection standards – plan of action for scour critical bridges, FHWA Bridge Technology Memorandum
Gordenker A (2011) Bridges with names, Japan Times, 17.5.2011, http://www.japantimes.co.jp
Hannawald F, Reintjes KH, Graße W (2003) Measurement-based assessment of the service behaviour of a steel composite motorway bridge. Stahlbau 72(7):507–516
Harik IE, Shaaban AM, Gesund H, Valli YS & Wang ST (1990) United States bridge failures
Heinrich B (1983) Brücken – Vom Balken zum Bogen. Rowohlt Taschenbuch Verlag GmbH, Hamburg
Hersi M (2009) Analysis of Bridge Failure in United States (2000–2008), Ohio State University, Master of Science Thesis
Hingorani R (2021) Personal communication
Hingorani T, Tanner P, Proske D, Syrkov A (2022) Development of Models for Estimation of Consequences to Persons due to Bridge Failures, IABSE TG 1.5, subchapter 3.3, Technical Report
Imam BM, Chryssanthopoulos MK (2012) Causes and consequences of metallic bridge failures. Struct Eng Int 22(1):93–98
Imhof D (2004) Risk Assessment of Existing Bridge Structures, University of Cambridge, Dissertation, Kings College, December 2004
Knoll F (1965) Grundsätzliches zur Sicherheit der Tragwerke, Promotion 3701, ETH Zürich, Zürich
Kroker H (2013) Deutschlands Brücken vor dem Kollapse. Die Welt, 3.6.2013. https://www.welt.de/116759711

Lee FZ, Lai JS, Lin YB, Liu X, Chang KC, Lin CF, Chang CC (2019) Monitoring and Simulation of Bridge Pier Scour Depth, Scour and Erosion, Keh-Chia, Taylor & Francis, pp 633
Lee GC, Mohan SB, Huang C & Fard BN (2013a) A Study of U.S. Bridge Failures (1980–2012), Technical Report MCEER-13–0008, June 15 2013, University at Buffalo, State University of New York
Lee GC, Qi J, Huang C (2013b) Development of a Database Framework for Modelling Damaged Bridges, Technical Report MCEER-13-0009, Jun 16 2013. University at Buffalo, State University of New York
Macbeth A (2018) Thousands of Italian Bridges will be in Crisis in the next 20 years. https://www.thelocal.it/20180828/thousands-of-italian-bridges-will-be-in-crisis-in-the-next-20-years
McLinn J (2010) Major Bridge Collapses in the US and Around the World, IEEE Reliability Society 2009 Annual Technology Report. IEEE Transactions on Reliability 59(3):5
Melville BW, Sutherland AJ (1988) Design method for local scour at bridge piers. J Hydraul Eng 114(10):1210–1226
Menzies JB (1996) Bridge Failures, Hazards and Societal Risk, International Symposium on the Safety of Bridges, July 1996. London
MLIT (2005) The maintenance of national road network in Japan, Ministry of Land, Infrastructure, Transport and Tourism, http://www.mlit.go.jp/road/road_e/03key_challenges/1-2-2.pdf
Montalvo C, Cook W (2017) Validating Common Collapse Conjectures in the U.S. Bridges, 11th International Bridge and Structure Management Conference, April 25–27 2017, Mesa, Arizona, 13
Naumann J (2002) Aktuelle Schwerpunkte im Brückenbau, 12th Dresden Bridge Construction Symposium, Chair of Solid Construction and Friends of Civil Engineering e.V., pp 43–5
Oliveira CBL, Greco M, Bittencourt TN (2019) Analysis of the Brazilian federal bridge inventory. IBRACON Struct Mater J 12(1):1–13
Pircher M, Kammersberger A, Lechner B, Mariani O (2009) Damage of a slackly reinforced concrete bridge by traffic loading. Beton- Und Stahlbetonbau 104(3):154–163
Proske D (2009) Catalogue of Risks, Springer: Heidelberg, New York
Proske D (2017a) Comparison of failure probability and failure frequency of bridges. Bautechnik 94(7):419–429
Proske D (2017b) Comparison of Bridge Collapse Frequencies with Failure Probabilities, Proceedings of the 15th International Probabilistic Workshop, Dresden, TUDpress, Eds. M. Voigt, D. Proske, W. Graf, M. Beer, U. Häussler-Combe, P. Voigt, September 2017, pp 15–23
Proske D (2018a) Versagenshäufigkeit und Versagenswahrscheinlichkeit von Brücken, Proceedings of the 28th Dresden Bridge Building Symposium, Dresden University of Technology, Dresden, pp 189–199
Proske D (2018b) Bridge Collapse Frequencies versus Failure Probabilities, Springer, Cham
Proske D (2019a) Comparison of the collapse frequency and the probability of failure of bridges. Proceedings of the Institution of Civil Engineers – Bridge Engineering 172(1):27–40
Proske D (2019b) Comparison of Frequencies and Probabilities of Failure in Engineering Sciences, Proceedings of the 29th European Safety and Reliability Conference, Edited by Michael Beer and Enrico Zio, Research Publishing, Singapore, p 2040–2044. https://doi.org/10.3850/978-981-11-2724-30816-cd
Proske D (2019c) The 30-year cycle of bridge collapses and technological and demographic change in civil engineering. Civil Engineer 94(9):343–352
Proske D (2020a) Extended comparison of failure probability and frequency of nuclear power plants, bridges, dams and tunnels. Civil Engineer 95(9):308–317

Proske D (2020b) On the consideration of hypothetical casualty figures in life cycle cost calculations of bridges. Beton- und Stahlbetonbau 115(6):459–46

Proske D (2020c) Fatalities due to bridge collapse. Proc Inst Civ Eng – Bridge Eng 173(4):260–267

Proske D (2020d) The global health burden of structural failure. Civil Eng 97(4):233–242

Proske D, van Gelder P (2009) Safety of historical Stone arch Bridges. Springer, Heidelberg

Schaap HS, Caner A (2021) Bridge collapses in Turkey: causes and remedies. Struct Infrastruct Eng. https://doi.org/10.1080/15732479.2020.1867198

Scheer J (2000) Failure of structures: Vol 1: Bridges. Ernst und Sohn Publishing House: Berlin

Sharma S & Mohan S (2011) Status of Bridge Failures in the United States (1800–2009). TRB 90th Annual Meeting: Transportation, Liveability, and Economic Development in a Changing World, Washington D.C

Sharma S (2010) A Comprehensive Study on Bridge Failures in the United States and their Comparison with Bridge Failure in Other Countries, MSc Project, Department of Civil, Structural and Environmental Engineering, University of Buffalo: Buffalo

Smith DW (1976) Bridge failures.Proc Inst Civ Eng 60(3):367–382

Spector A, Gifford D (1986) A computer science perspective of bridge design. Communication of the ACM 29(4):268–283

Statista (2020a) Number of road bridges in China from 2010 to 2019. https://www.statista.com/statistics/258358/number-of-road-bridges-in-china/

Statista (2020b) Number of road bridges in South Korea from 2009 to 2019. https://www.statista.com/statistics/1052665/south-korea-number-of-road-bridges/

Statistics Canada (2018) Canada's Core Public Infrastructure Survey: Roads, bridges and tunnels, 2016, 24 August 2018

Syrkov A (2020) Personal communication

Taricska MR (2014) An Analysis of Recent Bridge failures in the United States (2000–2012). Thesis, The Ohio State University, MSc

Tweed MH (1969) A summary and analysis of bridge failures, Master Thesis, Iowa State University, Ames

UIC (2005) International Union of Railways: Improving Assessment, Optimization of Maintenance and Development of Database for Masonry Arch Bridges. http://orisoft.pmmf.hu/masonry/

Vogel T, Zwicky D, Joray D, Diggelmann M & Hoj NP (2009) Tragsicherheit der bestehenden Kunstbauten, Sicherheit des Verkehrsssystems Strasse und dessen Kunstbauten, Federal Roads Office (ASTRA), December 2009, Bern

Wang Z, Lee GC (2009) A comparative study of bridge damage due to the Wenchuan, Northridge, Loma Prieta and San Fernando earthquakes. Earthq Eng Eng Vib 8(2):251–261

Wardhana K & Hadipriono FC (2003) Analysis of Recent Bridge Failures in the United States. Journal of Performance of Constructed Facilities, ASCE, August 2003, pp 144–150

Weber WK (1999) The arched railway bridge with an opening, Dissertation, Technical University of Munich, Munich

Wikipedia (2017) List of bridge failures. https://en.wikipedia.org/wiki/List_of_bridge_failures

Woodward RJ, Kschner R, Cremona C, Cullington D (1999) Review of current procedures for assessing load carrying capacity – Status C, BRIME PL97–2220. January 1999

Wolhuter KM (2015) Geometric design of roads handbook. CRP Press, Taylor & Francis Group, Boca Raton

Xu FY, Zhang MJ, Wang L, Zhang JR (2016) Recent highway bridge collapses in China: review and discussion. J Perform Constr Facil 30(5):8

Yan BF, Shao XD (2008) Application of China Bridge Management System in Qinyuan City. Bridge Maintenance, Safety, Management, Health Monitoring and Informatics - KOH & FRANGOPOL (eds), Taylor & Francis Group, London, pp 2675–2682

Zerna W (1983) Basis of current safety practice in civil engineering. In: Hartwig S (ed) Major technical hazards: risk analysis and safety issues. Springer, Berlin, pp 99–109

Zhang XG, Liu G, Ma JH, Wu HB & Wu WM (2014) Design concept and approach on sustainable development of bridge engineering, Bridge Maintenance, Safety, Management and Life Extension, Chen, Frangopol and Ruan (eds), CRC Press, London, pp 1831–1838

4 Dams

4.1 Introduction

Dams form an essential part of the infrastructure in developing and developed countries. They can be part of the roadway to meet the required gradients or act as a barrier structure or impoundment.

As hydraulic engineering structures, they serve various purposes, such as ensuring water supply for households and industry, flood protection, as part of irrigation systems, improving navigability of water bodies, as recreational areas, for fish farming, as sedimentation control, for power generation or several functions at the same time. Dams form a core element in water management, as do bridges in landbased transport systems. (ICOLD 2018a; DeNeale et al. 2019).

The first man-made dams were probably built more than 5,000 years ago in Mesopotamia and Egypt (ICOLD 2013; Yang et al. 1999; White 2016; Jansen 1983). Some dams with an age of 3,000 years are still in operation today (Water Technology 2013). Dam building probably became more important with the strong growth of transportation in the eighteenth century and industrialisation. Another maximum value of dam building probably occurred in Switzerland and the USA at the beginning of the twentieth century.

4.2 Definition of Dams

In this chapter, the terms dams, barrages, dikes, and dam walls are used.

A dam is an artificially constructed earth or stone structure, usually elongated and tapering towards the top (Seidel 2006). Naturally formed dams are not considered

D. Proske, *The Collapse Frequency of Structures*,
https://doi.org/10.1007/978-3-030-97247-9_4

here. Dams, as water-retaining structures, usually have a watertight core (Swiss Dams Committee 2021).

Concrete or masonry dams are artificial structures with a slimmer cross-section than embankment dams. (Swiss Dams Committee 2021).

According to DIN 19 700-11 (2004), dams are *"impoundments that seal off the cross-section of the valley beyond the cross-section of the impounded watercourse."* The Saxon Water Act (SächsGVBl 2016) defines dams in § 65, paragraph 1 as *"installations for the temporary or permanent impounding of a watercourse and for the storage of water, in which the height of the dam structure from the lowest air-side ground point at the dam structure to the crown is more than 5 m and the maximum permissible usable space comprises a volume of more than 100,000 m^3."*

4.3 Stock of Dams

The total number of dams worldwide is estimated at around 800,000 to 950,000 (Proske 2020b). Duffey and Saull (2003) estimate that there are about 80,000 dams in the USA alone. This estimate is also found, for example, in FEMA (2016).

According to Deangeli et al (2009), the global number of large dams was about 45,000 in the year 2000, according to International Rivers (2018) it is in the range of about 40,000, while ICOLD (2018b) estimates about 36,000 large dams and Zhang et al (2016) estimate about 25,000 large dams. According to Deangeli et al (2009), over 50% of large dams are in China and over 75% are in Asia. A large dam has a height of at least 15 m from base to crest or has a height of 5 to 15 m but with a reservoir of more than 3 million m^3 (ICOLD 2011). The BWG (2003) recognises the classification of large dams with a height of 40 m or 10 m and a storage volume of at least 1 million m^3, medium dams with a height of 25 m or a minimum height of 5 m to 15 m and staggered storage volume, and small dams.

The different definitions are again shown clearly, but here in relation to the crown height and not to the span as with bridges. But as with bridges, these different definitions naturally have an impact on the counting of the structures.

For example, the number of particularly significant dams is only in the region of 300 (International Rivers 2018).

4.4 Calculation of Collapse Frequencies

4.4.1 Introduction

There are numerous publications on dam failure and statistical evaluation (Baecher et al. 1980; Jansen 1983; Ott et al. 1984; ICOLD 1995; Tatolovich 1998; Ellingwood 1999; Lemperiere 1999; Foster et al. 2000; Cenderelli 2000; Ebi 2007; Ferrante et al. 2013;

NRC 2013; US DoI 2014, 2015; LePoudre 2015; Zhang et al. 2016; Wikipedia 2018; ASDOS 2018; International Rivers 2018; Proske 2018; DeNeale et al. 2019). Examples of dam failure are also mentioned in Miedema (2009).

For Switzerland, Lombardi Engineering (2014) has produced studies that also consider the trend in collapse frequencies. Something comparable for the USA can be found in Duffey and Saull (2003) and Ferrante et al. (2013). Penman (1986) shows the growth curves for the number of dams and the height of dam's worldwide from 1800 to 1985. While Lombardi Engineering (2014) and Duffey and Saull (2003) see the peak of dam construction at the beginning of the twentieth century for Switzerland and the USA, Penman (1986) assumes a parallel growth of dams and world population.

The risks of dam failure as part of the risks of electricity supply have also been studied intensively (Hauptmanns et al 1987; Holder 2016; Bauer et al. 2017; European Commission 1995) especially in Switzerland (Kalinina et al. 2016a, b; Kalinina et al 2017).

4.4.2 Central Estimator

Figure 4.1 shows the mean collapse frequencies of dams according to different publications and for different time periods. The averaging includes certain assumptions. Trend lines and results of probabilistic calculations are also integrated into the figure.

First of all, it can be seen that the dams of hydropower stations are significantly safer than all dams. The difference is at least one order of magnitude. It should be noted again that the data include a wide variety of dams and reservoirs.

Despite the large dispersion, a falling trend is recognisable from 1930 onwards, at least for the Swiss dams (blue line). Such a falling trend was also found for the USA (Baecher et al. 1980; Duffey and Saull 2003; Ferrante et al. 2013).

With country-specific classification of dams, very high collapse frequencies are sometimes observed, e.g. for China. This is also known for particular types of dams. For example, Azam and Li (2010) state a collapse frequency of 10^{-2} and LePoudre (2015) of 10^{-3} per year for retaining dams of mines. The data for China in the range of 10^{-2} and 10^{-1} per year fit into these publications but are interpreted here as outliers.

4.4.3 Measure of Deviation

In addition to determining mean collapse frequencies based on different data sets and publications, one can also determine the dispersion of these data. This applies to both aleatory uncertainty and epistemic certainty. The latter was included in Table 4.1 as the confidence interval of the mean, the former was reported as the standard deviation. The aleatory uncertainty is significantly lower for dams than for bridges, both for the observed collapse frequency and for the calculated failure probability. However, the

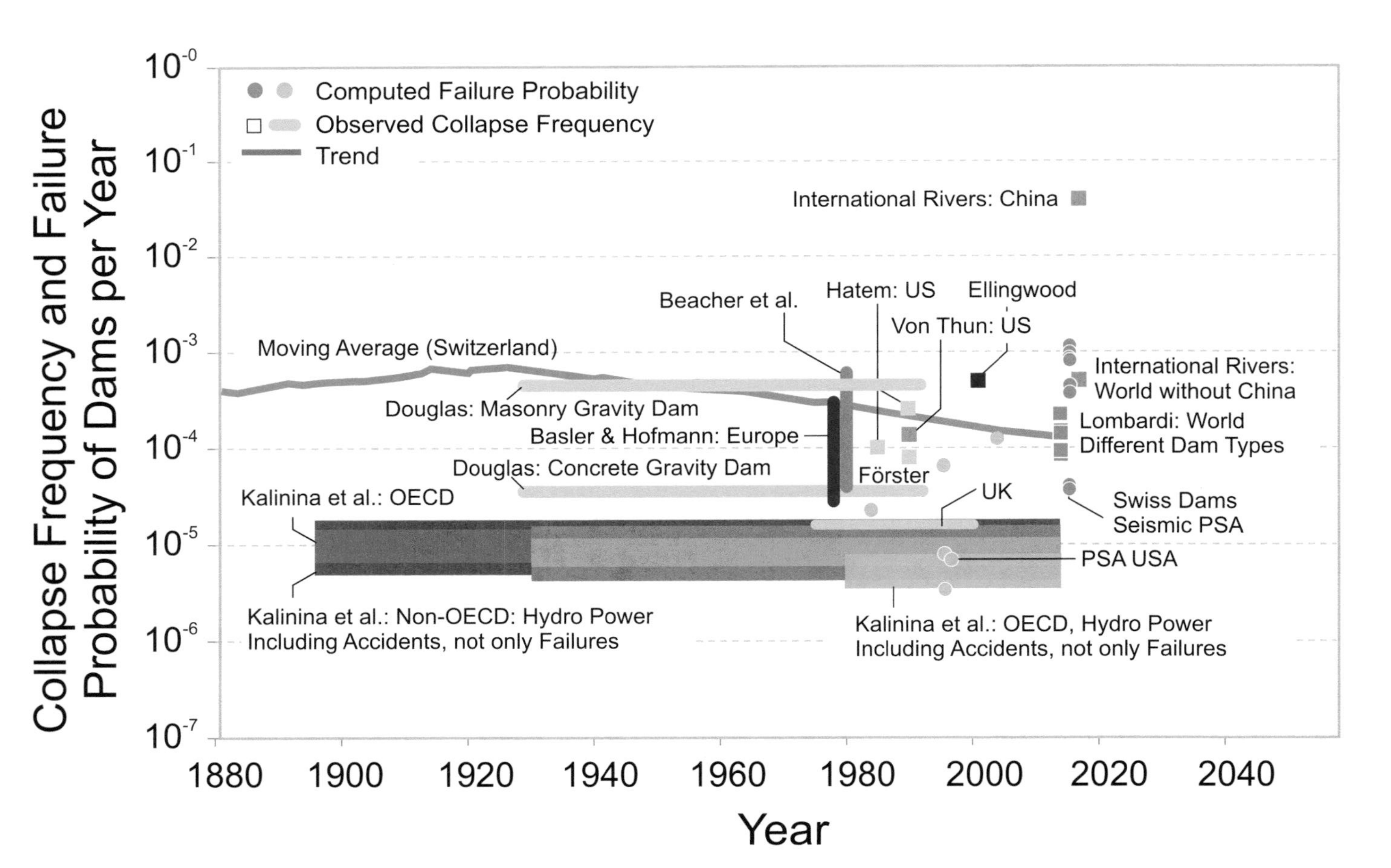

Fig. 4.1 Collapse frequencies and failure probabilities of dams (Proske 2020a, 2018)

Table 4.1 Mean values including confidence intervals and standard deviations (Proske 2020a)

Probabilistic calculations		Observed frequency		Target value per year
Standard deviation	Mean values incl. confidence interval	Standard deviation	Mean values incl. confidence interval	
3.46	3.46×10^{-4} (3.40×10^{-4}, 3.51×10^{-4})	2.72	3.02×10^{-4} (2.98×10^{-4}, 3.06×10^{-4})	10^{-6}

sample size for bridges was significantly larger for the nineteenth century (see Sect. 3.6). This at least partly explains the larger scatter.

4.4.4 Trend

As a function type of the temporal trend of the mean observed collapse frequencies, an exponentially decreasing trend is found in the literature (Proske 2018). The decadic logarithm of the mean change per century is about 1.04 (Proske 2020a). This corresponds to a reduction in collapse frequency by a factor of 11 per century.

4.5 Calculation of Failure Probabilities

The selected probabilistic calculations mainly refer to two publications: A collection of seismic fragility calculations of almost 20 Swiss dams and six probabilistic calculations of American dams. In addition, further calculations are available in Hill et al. (2013), Tekie and Ellingwood (2003) or references to probabilistic calculations in Stedinger et al. (1996), which, however, were not used.

4.6 Comparison

Figure 4.2 shows a comparison of the frequency distribution of collapse frequencies and failure probabilities. For dams, the sample size of the collapse frequency studies, and the probabilistic calculations practically match.

Figure 4.2 shows a very good match of the shape of the frequency distribution. This also fits with the comparable means and standard deviations in Table 4.1. However, the high value for China clearly stands out in Fig. 4.2.

Figure 4.3 shows a comparison of observed and calculated failure values according to Tatolovich (1998). This figure confirms the relatively good agreement in Fig. 4.2.

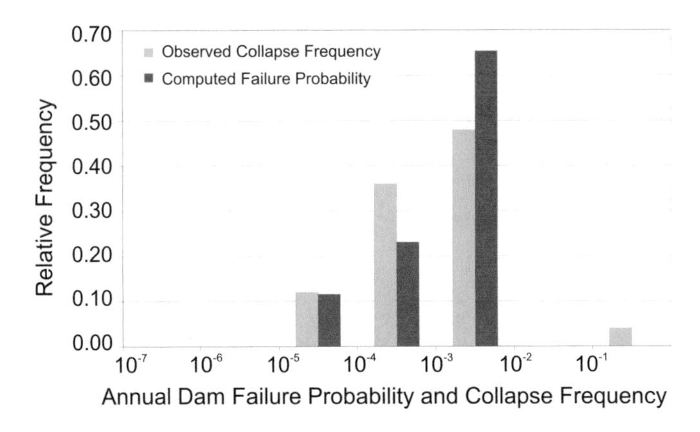

Fig. 4.2 Histogram of the calculated failure probabilities and the observed collapse frequencies for dams (Proske 2020b)

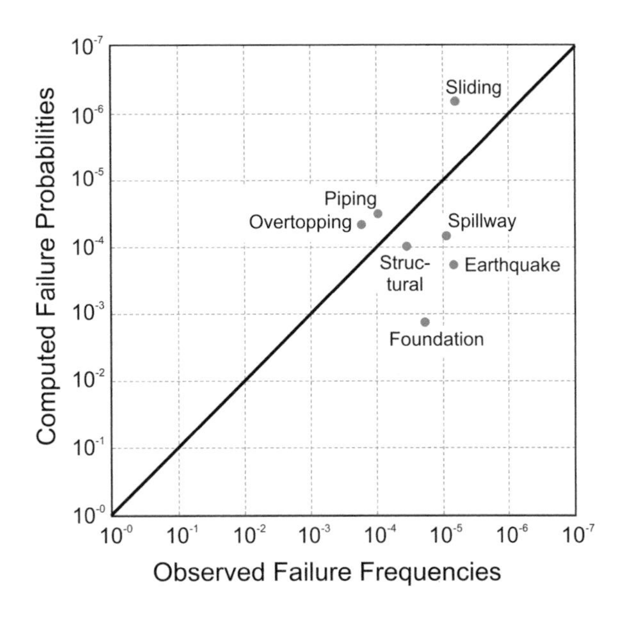

Fig. 4.3 Comparison of the calculated failure probabilities and observed failure frequencies according to Tatolovich (1998). In contrast to the other notation, the term failure frequency was used here because it was used by Tatolovich (1998)

4.7 Causes of Collapse

Just as with bridge collapses, the causes of dam failures can be assigned to different thematic groups, such as various human errors or actions. Duricic (2014) summarises the causes as follows: *"The main reasons for [dam] failures are inconsistency between design and reality, natural processes, earthquakes and deliberate human actions."*

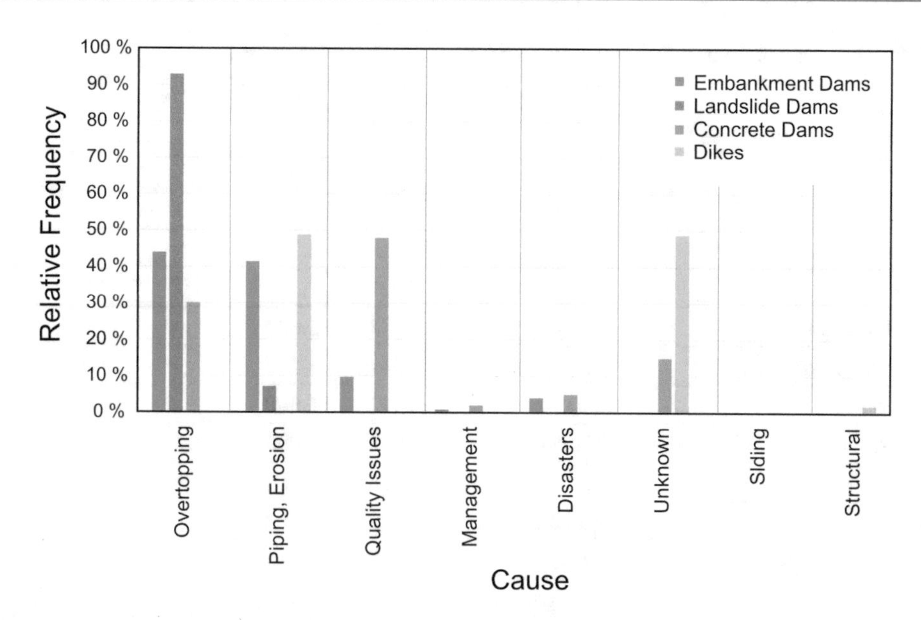

Fig. 4.4 Causes of dam failure according to Zhang et al. (2016)

However, the exact classification of the causes of collapse depends on the respective type of dam. Both in defining the terms dams, barrages, dikes, and dam walls and in determining the number of dams, it became apparent that a large number of different construction forms and types exist. Failure has different causes depending on the type of construction.

Figure 4.3 has already named various causes of failure. Figure 4.4 assigns the causes to the construction types. For comparison, natural dams were taken into account. For man-made dams, overtopping and erosion predominate as causes. This is also consistent with Zhang et al. (2007) and DeNeale et al. (2019).

4.8 Mortality and Casualty Figures

Of the engineering products assessed in this book, dams have the largest single loss event with nearly 200,000 direct and indirect fatalities (Banqiao and Shimantan Dam Failure 1975) (Proske 2009). Theoretically, dams can endanger many more people. In the Oroville Dam spillway failure in the US, 180,000 people were evacuated (Williamson 2017). During the exceptionally high-water level at Kakhovskaya Dam in Ukraine in February 1996, more than 500,000 people were at risk of dam failure (International Rivers 2018). Vrijling et al. (2001) considers one million affected people within a polder in the Netherlands. In the event of a failure of the Sihl-Dam in Switzerland, parts of Zurich would be flooded to a height of 8 m with devastating financial consequences.

Table 4.2 Dam failures with fatality numbers (Proske 2018; Williamson 2017)

Dam	Country	Year	Approximate number of fatalities
Banqiao and Shimantan Dams	China	1975	175,000
Machchu-2 Dam	India	1979	5,000
South Fork Dam	USA	1889	2,200
Vajont Dam	Italy	1963	2,000
Sempor Dam	Indonesia	1967	2,000
Möhne Dam	Germany	1943	1,600
Kurenivka landslide	Soviet Union	1961	1,500
Tigra Dam	India	1917	1,000
Panshet Dam	India	1961	1,000
Iruka Lake Dam	Japan	1868	950
Puentes Dam	Spain	1802	610
St-Francis Dam	USA	1928	600
Vratsa	Bulgaria	1966	600
Malpasset Dam	France	1959	420
Vega de Tera	Spain	1959	400
Gleno Dam	Italy	1923	350
Val di Stava Dam	Italy	1985	270
Koshi Barrage	Nepal	2008	250
Rapid City	USA	1972	250
Dale Dike Reservoir	Great Britain	1984	240
Qued-Fergoug	Algeria	1881	200
Bouzey	France	1895	100
Austin	USA	1910	100

Based on Table 4.2, a mean number of victims per year due to dam failure of about 1,000 can be determined, with the above-mentioned single event dominating the mean value.

Graham (1999) provides a table of dam failures in the USA with fatalities for the period 1960 to 1998. NN (2009) has provided a list of fatal and non-fatal dam failures and "near failures" for the period 1875 to 2008.

According to Guha-Sapir et al. (2016), the number of annual global fatalities from surface water flooding is around 5,000. Jonkman (2005, 2007) cites around 175,000 fatalities from global flooding for the period from 1975 to 2001. This corresponds to a worldwide annual mean number of 7,000 victims. The author does not know what proportion of this number can be attributed to dam failure, but the average number for dams cannot exceed this value.

Individual events can also have a significant impact on fatality figures, such as the great Southeast Asian tsunami triggered by the 2004 Sumatra–Andaman earthquake, which claimed over 240,000 lives, or the flooding triggered by the 1970 Bhola cyclone in East Pakistan, which claimed well over 300,000 lives (Frank and Husain 1971). In fact, flooding is one of the biggest disasters. However, in the case of bridges, floods do not usually lead to increased mortalities because the bridges or the routes are closed beforehand.

A figure on dam failure mortality can be found, for example, in Lind and Hartford (1999). They cite a mortality of 2.5×10^{-5} for dam failures resulting in death in the USA.

In addition to mortality data, there are numerous publications with *f-N* or *F-N* diagrams for dam failures due to the possible maximum fatalities (NRC 1975; BZS 1995; Bea 1998; BABS 2003; Cook 2004; Eckle et al. 2011; Pohl and Bornschein 2011; US DoI 2015; DeNeale et al. 2019). Figures 4.5 and 4.6 show two examples.

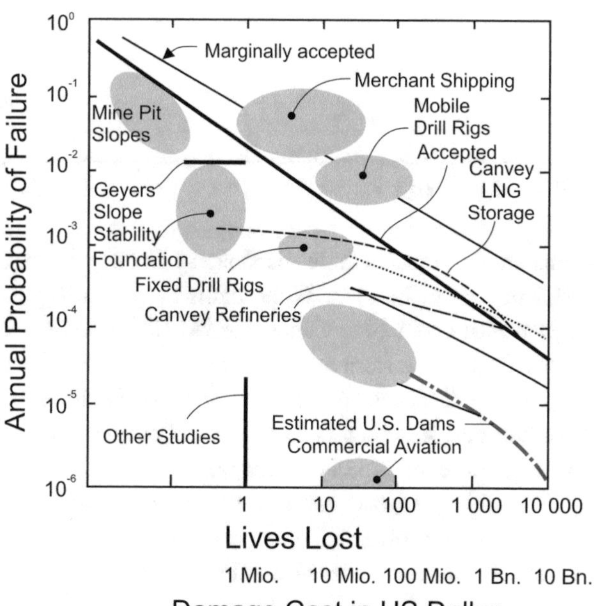

Fig. 4.5 *f/F-N* diagram with dam failure and other technologies for comparison (Bea 1998)

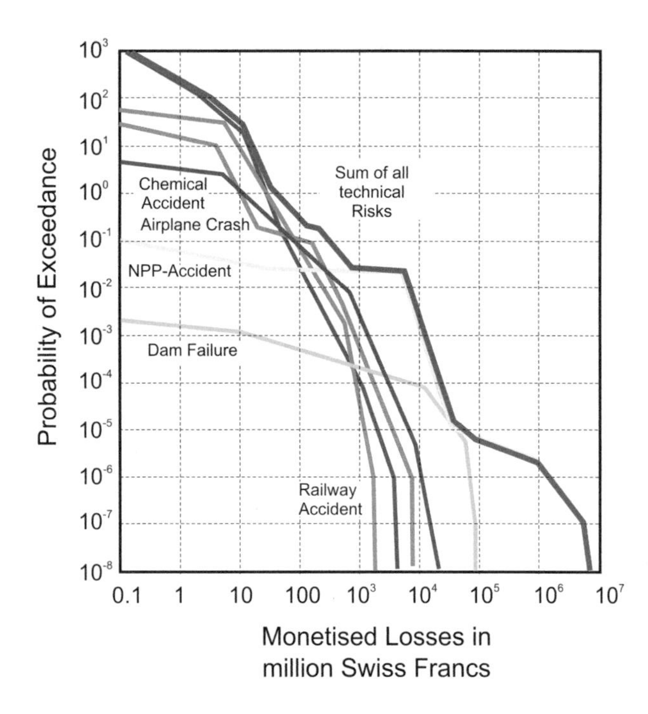

Fig. 4.6 *F-N* diagram for Swiss dams and other technical risks for comparison (BABS 2003; BZS 1995)

4.9 Summary

Dams show a considerable dispersion of collapse frequencies with respect to the type and function of the dam. While dams as part of hydropower stations show a collapse frequency of about 10^{-5}, values close to 10^{-1} have been observed for all types of dams in China. This is a concern not only from a collapse frequency perspective, but also because dam failure can cause extreme numbers of fatalities. There are practically no types of structures and, in fact, no other technical products whose individual failure causes fatalities in the order of 175,000 to 200,000.

From this perspective, the observed collapse frequency appears surprisingly high compared to the collapse frequency of bridges.

This chapter is an extension of Proske (2018).

References

ASDOS (2018) Failure and Incidents at Dams. https://damsafety.org/dam-failures, Association of State Dam Safety Officials, 2018

Azam S, Li Q (2010) Tailings Dam Failures: A Review of the Last One Hundred Years, Geotechnical News, December 2010, pp 50–53.

FOCP (2003) CATARISK - Disasters and emergencies in Switzerland, a risk assessment from the perspective of civil protection, Federal Office for Civil Protection, Bern

Baecher GB, Pate ME, De Neufville R (1980) Risk of dam failure in benefit-cost analysis. Water Resour Res 16(3):449–456

Bauer C, Hirschberg S (eds.), Y Bäuerle, S Biollaz, A Calbry-Muzyka, B Cox, T Heck, M Lehnert, A Meier, H-M Prasser, W Schenler, K Treyer, F Vogel, HC Wieckert, X Zhang, M Zimmermann, V Burg, G Bowman, M Erni, M Saar, MQ Tran (2017) Potentials, costs and environmental assessment of electricity generation technologies. PSI, WSL, ETHZ, EPFL. Paul Scherrer Institute, Villigen PSI

Bea RG (1998) Oceanographic and reliability characteristics of a platform in the Mississippi River Delta. Journal of Geotechnical and Geo-Environmental Engineering, ASCE 124(8):779–786

BWG (2003) Safety of dams. Basic document for the verification of seismic safety, Federal Office for Water and Geology, Bern

BZS (1995) Bundesamt für Zivilschutz. KATANOS – Disasters and Emergencies in Switzerland, a comparative overview, Bern

Cenderelli DA (2000) Floods from natural and artificial dam failures. Ed. Wohl, Inland Flood Hazards: human, riparian, and aquatic communities, Cambridge University Press, New York, pp 73–103

Cook W (2004) Bridge Failure Rates. Diss, Utah State University Logan, Consequences and Predictive Trends

Deangeli C, Giani GP, Chiaia B & Fantilli AP (2009) Dam failures, WIT Transactions on State of the Art in Science and Engineering, Vol 36, WIT Press, pp 1–50

DeNeale ST, Baecher GB, Stewart KM, Smith ED, Watson DB (2019) Current State-of-Practice in Dam Safety Risk Assessment, Oak Ridge National Laboratory, US NRC, ORNL/TM-2019/1069

DIN 19 700–11 (2004) Dams - Part 11: Dams, July 2004

Duffey, R.B. & Saull, J. W. (2003) Know the Risk: Learning form Errors and Accidents: Safety and Risk in Today's Technology, Butterworth-Heinemann

Duricic J (2014) Dam Safety Concepts, Dissertation, TU Delft, Delft

Ebi P (2007) Risk Analysis for Hydropower, Paul Scherrer Institute, January 2007

Eckle P, Cazzoli E, Burgherr P, Hirschberg S (2011) Analysis of terrorism risk for energy installations, Confidential Report, SECURE Deliverable No 5.7.2b, SECURE project Security of Energy Considering its Uncertainty, Risk and Economic Implications, Brussels, Belgium

Ellingwood BR (1999) Probability-based structural design: prospects for acceptable risk bases, Application of Statistics and Probability (ICASP 8). Sydney 1:11–18

European Commission (1995) Wind & Hydro, Vol. 6, ExternE Externalities, Science Research Development, Studies

FEMA (2015) Federal Guidelines for Dam Safety Risk Management. Federal Emergency Management Agency, Washington DC, p 332

FEMA (2016) Dam Safety Program, Federal Emergency Management Agency. www.fema.gov/hazards/damsafety

Ferrante F, Bensi M, Mitman J (2013) Uncertainty Analysis for Large Dam Failure Frequencies Based on Historical Data, NRC, ADAMS Accession No. ML1398A170

Foster M, Fell R, Spannagle M (2000) The statistics of embankment dam failures and accidents. Can Geotech J 37(5):1000–1024

Frank NL, Husain SA (1971) The deadliest cyclone in history? Bull Am Meteor Soc 52(6):438–445

Graham WJ (1999) A Procedure for Estimating Loss of Life Caused by Dam Failure, US DoI, Dam Safety Office, September 1999

Guha-Sapir D, Hoyois P, Wallemacq P, Below R (2016) Annual Disaster Statistical Review 2016: The Numbers and Trends. Centre for Research on the Epidemiology of Disasters, Brussels

Hauptmanns U, Werner W, Herttrich M (1987) Technical risks, determination and assessment. Springer Verlag, Berlin - Heidelberg GmbH

Hill P, Bowles D, Jorand P, Nathan R (2013) Estimating overall risk of dam failure: practical considerations in combining failure probabilities, ANCOLD 2013 Workshop, p 10

ICOLD (1995) International Commission on Large Dams, Bulletin 99. Dam Failures - Statistical Analysis, Paris

ICOLD (2011) International Commission on Large Dams, Constitution Statutes, July 2011, Paris

ICOLD (2013) International Commission on Large Dams, Bulletin 143: Historical review on ancient dams, Paris

ICOLD (2018a) International Commission on Large Dams: Role of Dams. http://www.icold-cigb

ICOLD (2018b) International Commission on Large Dams, Dams' safety is at the very origin of the foundation of ICOLD. www.icold-cigb.net

Holder H (2016) Energy Risk Assessment, Routledge, New York (1982, latest Edition)

International Rivers (2018) Damming Statistics. https://www.internationalrivers.org/damming-statistics

Jansen RB (1983) Dams and Public Safety, A Water Resources Technical Publication, U.S. Department of the Interior, Bureau of Reclamation, Denver

Jonkman SN (2005) Global perspectives on loss of human life caused by floods. Nat Hazards 34:151–175

Jonkman SN (2007) Loss of life estimation in flood risk assessment – Theory and applications. PhD thesis, Rijkswaterstaat - Delft Cluster, Delft

Kalinina A, Sacco T, Spada M, Burgherr P (2017) Risk Assessment for Dams of different Types and Purposes in OECD and non-OECD countries with a Focus on Time Trend Analysis, HYDRO 2017. Seville, Spain, p 11

Kalinina A, Spada M, Marelli S, Burgherr P, Sudret B (2016a) Uncertainties in the Risk Assessment of Hydropower Dams: State-of-the-Art and Outlook, Paul Scherrer Institute, ETH Zürich

Kalinina A, Spada M, Burgherr P, Marelli S, Sudret B (2016b) A Bayesian hierarchical modelling for hydropower risk assessment, Risk, Reliability and Safety: Innovating Theory and Practice - Walls, Revie & Bedford (eds.), Taylor and Francis Group, London 2017 (ESREL), pp 412–418.

Lemperiere F (1999) Risk analysis: what sort should be applied and to which dams? Hydropower & Dams, Issue 4:128–132

LePoudre DC (2015) Examples, Statistics and Failure modes of tailings dams and consequence of failure, REMTECH, 15 Oct. 2015, Presentation, 42 Slides

Lind N, Hartford D (1999) Probability of human instability in flooding: A hydrodynamic model, Application of Statistics and Probability (ICASP 8). Sydney 2:1151–1156

Lombardi Engineering Ltd (2014) Minusio, Switzerland

Miedema D (2009) Selected Case Histories of Dam Failures and Accidents Caused by Internal Erosion, https://www.usbr.gov/ssle/damsafety/TechDev/DSOTechDev/DSO-04-05.pdf

NN (2009) Dam Failures, Dam Incidents (Near Failures), Association of State Dam Safety Officials. www.damsafety.org

NRC (1975) An Assessment of Nuclear Risks in U.S. Commercial Nuclear Power Plants. WASH-1400, NUREG 75/014, US Nuclear Reactor Commission, October 1975, National Technical Information Service, Springfield

NRC (2013) Comparison of Approaches for Calculating a Jocassee Dam Failure Frequency, US Nuclear Reactor Commission

Ott KO, Hoffmann H-J, Oedekoven L. (1984) Statistical Trend Analysis of Dam Failures since 1850, Kernforschungszentrum Julich Gmbh (KFA), Jul-SPez-245

Penman ADM (1986) On the embankment dam. Geotechnique 36:301–348
Pohl R, Bornschein (2011) How safe is safe? Experiences in Dam Safety Policy, Valencia, p 6
Proske D (2009) Catalogue of risks. Springer, Berlin
Proske D (2018) Comparison of Large Dam Failure Frequencies with Failure Probabilities, Beton- und Stahlbetonbau, Vol 113, Special Issue (S2): 16th International Probabilistic Workshop, pp 2–6
Proske D (2020) Extended comparison of failure probability and frequency of nuclear power plants, bridges, dams and tunnels. Civil Engineer 95(9):308–317
Proske D (2020b) The global health burden of structural failure. Civil Eng 97(4):233–242
SächsGVBl (2016) Saxon Water Act of 12 July 2013 (SächsGVBl. p. 503), as last amended by Article 2 of the Act of 8 July 2016 (SächsGVBl. p. 287)
Swiss Dams Committee (2021) The different construction methods of dams. http://www.swissdams.ch
Seidel H (2006) Lexikon der Bautypen: Functions and Forms of Architecture, Reclam, Stuttgart Stedinger JR, Heath DC
Thomson K (1996) Risk Analysis for Dam Safety Evaluation: Hydrologic Risk, IWR Report 96-R-13
Tatolovich J (1998) Comparison of Failure Modes from Risk Assessment and Historical Data for Bureau of Reclamation Dams, January 1998, Materials Engineering and Research Laboratory, DSO-98–01, U.S. Department of the Interior, Bureau of Reclamation, Dam Safety Office
Tekie PB, Ellingwood BR (2003) Perspectives on probabilistic risk assessment of concrete gravity dams, Applications of Statistics and Probability in Civil Engineering, Der Kiureghian, Madanat & Pestana (Eds), Millpress, Rotterdam, pp 1725–1732
US DoI (2014) RCEM - Reclamation Consequence Estimation Methodology: Dam Failure and Flood Event Case History Compilation, February 2014. https://www.usbr.gov/ssle/damsafety/documents/RCEM-CaseHistories20140731.pdf
US DoI (2015) Best Practices in Dam and Levee Safety Risk Analysis, Part IX - Risk Assessment/Management, Version 4.0, July 2015
Vrijling JK, van Gelder PHAJM, Goossens LHJ, Voortman HG, Pandey MD (2001) A Framework for Risk criteria for critical Infrastructures: Fundamentals and Case Studies in the Netherlands, Proceedings of the 5th Conference on Technology, Policy and Innovation, "Critical Infrastructures", Delft, The Netherlands, June 26–29, 2001, Uitgeverrij Lemma BV
Water Technology (2013) The world's oldest dams still in use. http://www.water-technology.net/features/feature-the-worlds-oldest-dams-still-in-use/
White E (2016) A History of Dams: From Ancient Times to Today, Tata & Howard, May 17 2016. https://tataandhowardcom/2016/05/a-history-of-dams-from-ancient-times-to-today/
Wikipedia (2018) Dam failures. https://en.wikipedia.org/wiki/Dam_failure
Williamson T (2017) Historic dam failures and recent incidents, Presentation, ARUP
Yang H, Hayes M, Winzenread S, Okada K (1999) The History of Dams. https://watershed.ucdavis.edu/shed/lund/dams/Dam_History_Page/History.htm
Zhang LM, Peng M, Chang D, Xu Y (2016) Dam Failure Mechanisms and Risk Assessment, 17 June 2016
Zhang LM, Xu Y, Jia JS (2007) Analysis of earth dam failures – A database approach, SGSR2007 First International Symposium on Geotechnical Safety & Risk, Oct. 18–19 (2007) Shanghai. Tongji University, China, pp 293–302

Tunnel

5

5.1 Introduction

The same starting point applies to tunnels as to bridges with regard to the exponentially increasing volume of traffic. For this reason, numerous extraordinary tunnel construction projects have been planned and implemented in recent years, such as the Eurotunnel and the Gotthard Base Tunnel, or are still under construction, such as the Brenner Base Tunnel (Bergmeister 2007, 2010).

First tunnels were built at least 4,000 years ago. Early tunnels were built in Egypt, India, and China. The first tunnel structure whose engineer is known is the 1036 m long water tunnel of Eupalinos of Megara on Samos, Greece, built around 530 BC (ITA 2019; Sandström 1963).

5.2 Definition of Tunnels

According to ITA (2019), a tunnel is an artificial underground passage that is open at both ends. The Swiss Federal Roads Office (FEDRO 2014) recognises a minimum length of 300 m for the area between fully closed tunnel sections as a criterion for the application of certain procedures and rules.

The SIA 198 (2004) considers tunnels as "*Underground construction of great length and with a maximum slope of 20%.*"

DIN 1076 (1999) defines tunnels as "*structures serving road traffic, which lie below the surface of the ground or water and are constructed using a closed construction method or, in the case of an open construction method, are longer than 80 m*".

Striegler (1993) defines "*Tunnels are underground, secured, elongated cavities that serve traffic and emerge on both sides.*"

D. Proske, *The Collapse Frequency of Structures*,
https://doi.org/10.1007/978-3-030-97247-9_5

The regulation ABBV (2010) defines "*Tunnels are constructed using the closed (mining) or open construction method. Structures built using the cut-and-cover method are only considered to be tunnels above a certain length: for railways from 250 m, for roads from 80 m...*".

According to CD 352 (2020), "*a road tunnel is defined as a subsurface highway structure enclosed for a length of 150 m, or more, measured along the centre line of the soffit.*"

Sometimes also the tunnel bore area is used for tunnel definitions.

DIN 1076 (1999) also defines which parts belong to a tunnel. The Swiss Federal Railway considers different elements, such as the tunnel structure with vault and invert, portal wall with the adjacent wing walls and shafts and galleries, as far as they are connected to the main tube (EBP 2014). Tunnels are often divided into cut-and-cover tunnels and mining tunnels regarding their construction (Vogel et al. 2009) and their usage (Thewes and Maidl 2013; Maidl et al. 2014).

5.3 Stock of Tunnels

In recent years, the number of tunnels and tunnel construction projects has increased rapidly worldwide. In 2016, the annual growth was around 7% (ITA 2016, 2017a, see also 2017b, PR Newswire 2020, Schäfer 2019, GlobalData 2019).

The current global stock of tunnels is officially estimated at around 40,000 (ITA 2019). Table 5.1 shows a compilation of country-specific tunnel numbers. For some countries, such as Switzerland (STS 2020; SBB 2018, FEDRO 2016), detailed figures

Table 5.1 Stock of tunnels according to Spyridis and Proske (2021)

Country	Stock of tunnels
China	27,000
Japan	16,000
Norway	2,350
Korea	1,500
Germany	1,500
Switzerland	1,300
USA	850
Great Britain	600
Spain	420
Netherlands	60
Hong Kong	30
Sweden	35
Total	51,645

are available. For most countries, only limited information is available (DoT 2020; Wikipedia 2020; Statista 2020), making it difficult to estimate.

Spyridis and Proske (2021) therefore assume that there are more than 125,000 tunnels worldwide.

5.4 Calculation of Collapse Frequencies

5.4.1 Introduction

The evaluation of the collapse frequency of tunnels, in contrast to the other structure types, was carried out twice, once in Proske et al. (2019) and once in Spyridis and Proske (2021). In addition, there is an unpublished study to investigate the trend of collapse frequency.

Table 5.2 summarises the essential information on the samples of the two publications.

References with studies of observed tunnel collapse frequency are Seidenfuss (2006), Zhao (2009), Sousa (2010), Reiner (2011), Špačková et al. (2013), CEDD (2015) and Zhang et al. (2016).

5.4.2 Central Estimator

In the statistical evaluation of tunnel collapses, two special features must be considered:

- The majority of tunnel collapses (80–90%) occur during the construction phase. For comparison, it should be mentioned that the share of collapses during the construction period for bridges is about 30%.
- Many damages are related to accidents of transport vehicles inside the tunnel and subsequent fires (Ingason et al. 2015).

Figures 5.1 and 5.2 show the development of the collapse frequency of tunnels based on the evaluation of individual collapses and comparative values in the literature. Figure 5.1

Table 5.2 Comparison of the sample sizes of the two publications

	Proske et al. (2019)	Spyridis and Proske (2021)
Number of tunnel collapses	114	321
Proportion of collapses during the construction period	80%	92%
Number of probabilistic calculations	3	31

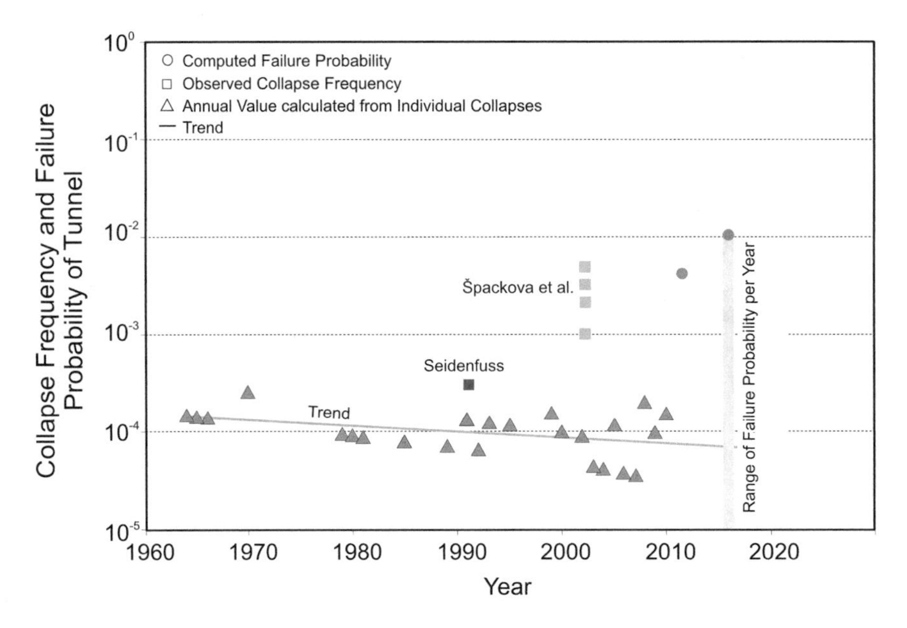

Fig. 5.1 Collapse frequencies and failure probabilities of tunnels (Proske et al. 2019; Proske 2019)

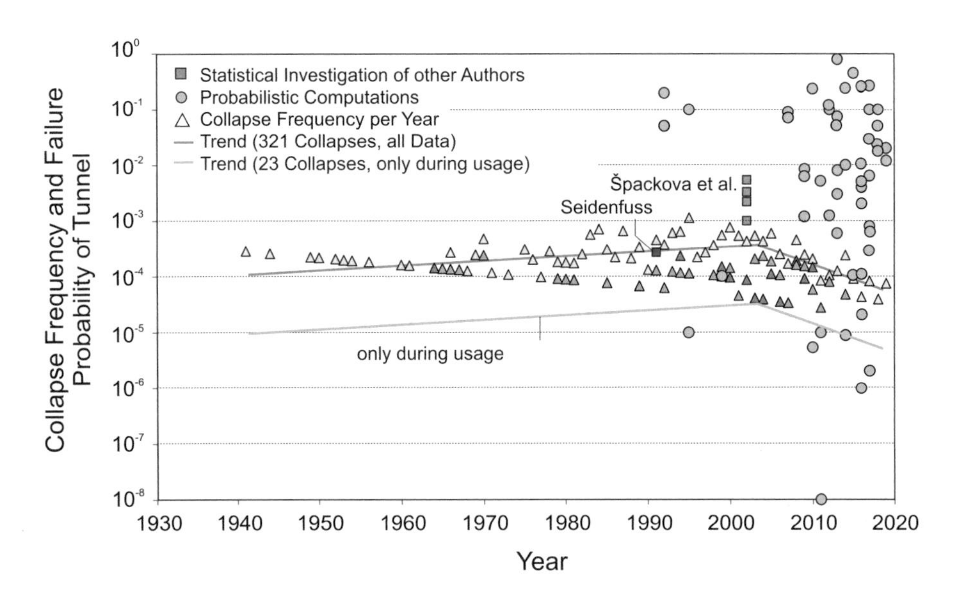

Fig. 5.2 Collapse frequencies and failure probabilities of tunnels (Spyridis and Proske 2021)

is based on Proske et al. (2019), Fig. 5.2 is based on Spyridis and Proske (2021). In Fig. 5.2 there is a trend line for collapses during construction and use and for collapses exclusively during use. The two figures allow not only the assessment of collapse frequency, but also an evaluation of the effects of limited sampling in the context of these studies. The different trend between the two figures and the larger calculated failure probabilities are clearly visible.

It is interesting to ask whether the peak around the year 2000 is a statistical artefact or a real phenomenon. Between 1994 and 2003, several major tunnel collapses occurred, such as the Munich Underground, the Great Belt Link, Heathrow Airport and the L.A. Metro Tunnel (1994–1995). In 2003, partly in response to these tunnel collapses, the Joint Code of Practice for Risk Management of Tunnel Works was introduced by the British Tunneling Society and the Construction Risk Insurers (BTS 2003, see also Athansopoulou et al. 2019). The International Tunnel Association adopted it presumably because of other tunnel disasters around the world (such as the Nicholson Highway collapse in 2004). The introduction of this document may have led to the reversal of the trend from 2000 onwards.

5.4.3 Measure of Deviation

In addition to determining mean collapse frequencies based on different data sets and publications, one can also determine the dispersion of these data, both for aleatory uncertainty and epistemic certainty. The latter was included in Table 5.3 as the confidence interval of the mean, the former was reported as the standard deviation. The aleatory uncertainty for the observed collapse frequency is approx. 2.5. For the calculated failure probability, the determination of a standard deviation based on the small data of

Table 5.3 Mean values including confidence intervals and standard deviations (Proske et al. 2019, updated with data from Spyridis and Proske 2021)

Reference	Probabilistic calculations		Observed frequency		Target value per year
	Standard deviation	Mean values including confidence interval	Standard deviation	Mean values including confidence interval	
Spyridis and Proske (2021)	2.97	9.07×10^{-3} (8.10×10^{-3}, 2.62×10^{-2})	2.43	2.36×10^{-4} (1.87×10^{-4}, 2.87×10^{-4})	10^{-6}
Proske et al. (2019)	Not meaningful	5.30×10^{-4} (3.43×10^{-4}, 8.18×10^{-4})	2.43	2.15×10^{-4} (2.13×10^{-4}, 2.18×10^{-4})	10^{-6}

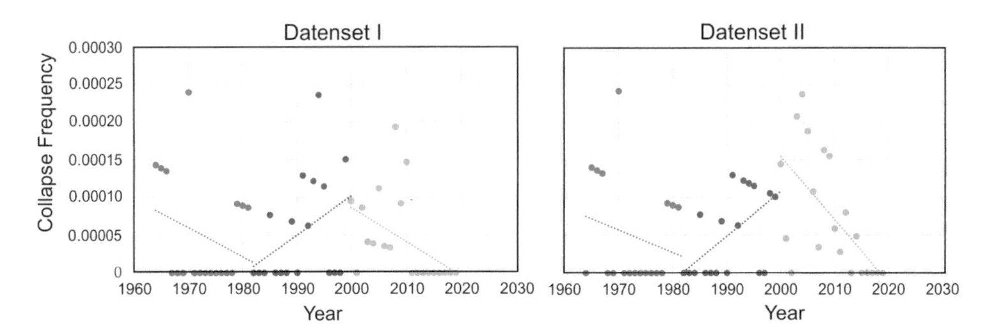

Fig. 5.3 Sequential regression for collapse frequency

Proske et al. (2019) was not meaningful. With the expanded sample size in the Spyridis and Proske (2021) study, a comparable standard deviation could be determined in the probabilistic calculations as in the observed collapse frequencies.

5.4.4 Trend

As a function type of the trend of the mean observed collapse frequencies, a linear decreasing trend is found in the literature (Proske et al. 2019). The decadic logarithm of the mean change per century is about 0.61. This would correspond to a reduction in collapse frequency per century by a factor of 4.

However, if one uses the data basis of Spyridis and Proske (2021), a wavy pattern may emerge. Regression analyses sometimes recommend a cosine function. A sequential linear regression is shown in Fig. 5.3. Data sets I and II correspond to the data from Reiner (2011) and CEDD (2015) as well as own extensions.

5.5 Calculation of Failure Probabilities

While the number of probabilistic calculations was very limited in the first study (3), it was significantly expanded in the second study with over 30 probabilistic calculations. The references for the probabilistic calculations are Kohno et al. (1992), Laso et al. (1995), Breitenbücher et al. (1999), Su et al. (2007), Mollon et al. (2009), Papaioannou et al. (2009), Bergmeister (2010), Goh and Hefney (2010), Li and Low (2010), Fortsakis et al. (2011), Lü et al. (2011), Goh and Zhang (2012), Lü et al. (2012), Langford and Diederichs (2013), Low and Einstein (2013), Lü et al. (2013), Gharouni-Nik et al. (2014), Spyridis (2014), Zhao et al. (2014), Miro et al. (2015), Bergmeister (2016), Johansson et al. (2016), Li et al. (2016), Spyridis et al. (2016), Wang et al. (2016), Zeng et al. (2016), Bjureland et al. (2017), Hamrouni et al. (2017), Lü et al. (2017), Liu and

Low (2017), Kroetz et al. (2018), Meschke et al. (2018), Yang et al. (2018a, 2018b) and Fuyong et al. (2019). Expanding the number of publications used will significantly improve the statistical significance of the evaluation results.

5.6 Comparison

Figures 5.4 and 5.5 show the frequency distribution of the observed collapse frequencies and the calculated failure probabilities. In addition, both figures show the changes based on the changes in sample sizes of the two studies.

While Fig. 5.4 allows practically no robust statements, Fig. 5.5 shows that the scatter of the calculations is significantly larger than the scatter of the observed collapse frequencies.

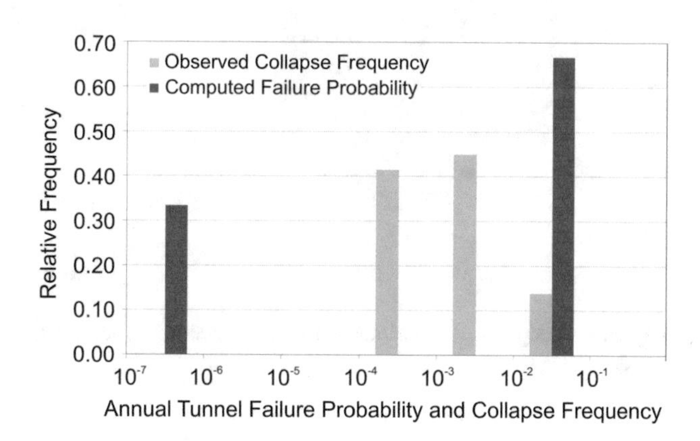

Fig. 5.4 Histogram of the calculated failure probabilities and the observed collapse frequencies of tunnels (Proske 2020 based on Proske et al. 2019)

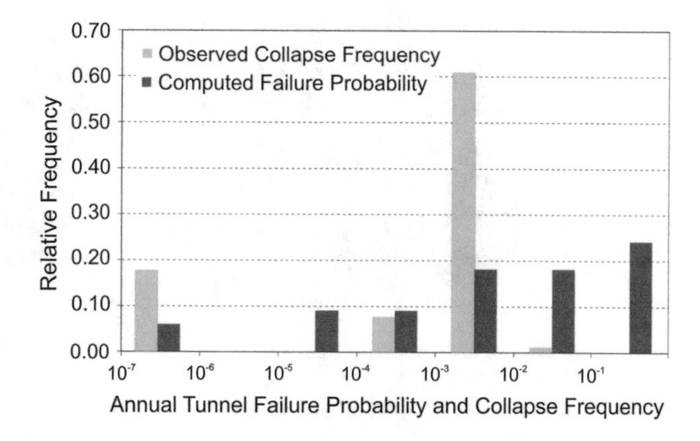

Fig. 5.5 Histogram of calculated failure probabilities and observed collapse frequencies of tunnels based on data from Spyridis and Proske (2021)

5.7 Causes of Collapse

In contrast to other types of structures, earthquakes and floods play a minor role as causes of tunnel collapses. The proportion of earthquake-induced collapses is less than 10% according to Fig. 5.6 on the right. Fires play a greater role, mostly in connection with accidents involving means of transport (Ingason et al. 2015).

However, failure is dominated by failure modes during construction (Proske et al. 2019). Due to this fact, the construction technology and the geological conditions play a dominant role. Therefore, in Fig. 5.7, the causes of collapse are assigned to the tunnel construction technology used.

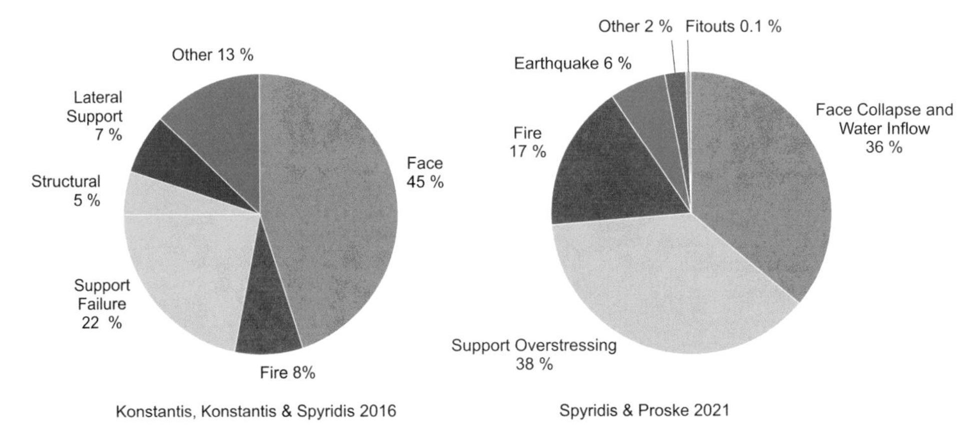

Fig. 5.6 Assignment of collapses to causes according to Konstantis et al. (2016) and Spyridis and Proske (2021)

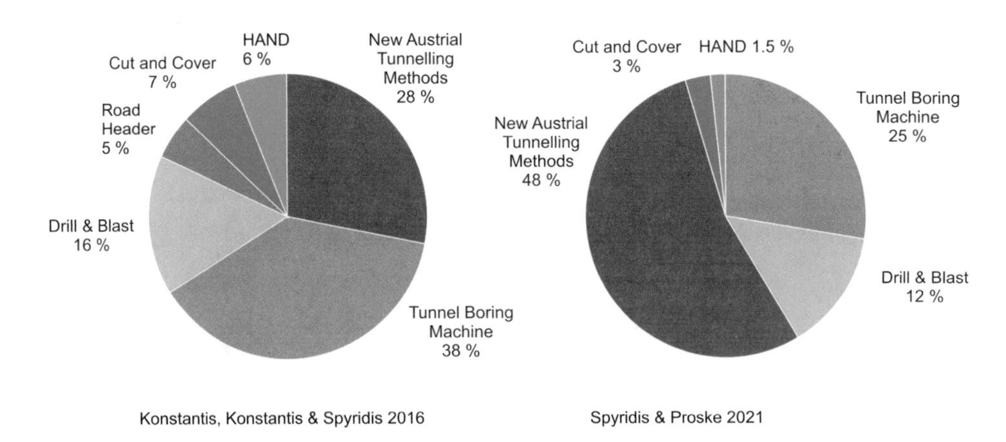

Fig. 5.7 Assignment of collapses to tunnelling technology according to Konstantis et al. (2016) and Spyridis and Proske (2021)

A qualitative assessment of collapse causes to actions can be found in Schubert and Faber (2008).

5.8 Mortality and Casualty Figures

The reports of fatalities in tunnels mostly concern accidents with means of transport (Ingason et al. 2015) and fire (Proske 2009). In addition, terrorist acts have also been described due to the limited escape possibilities in tunnels. According to Fig. 5.8, terrorist acts dominate the absolute numbers. A list of accidents with fatalities can also be found, for example, in Vogel et al. (2009). Overall, however, the number of fatalities is low.

According to Kröger and Høj (2000), the risk in road tunnels is not significantly greater than on the road (Fig. 5.9). Figure 5.10 shows the *F–N* target curves in different countries. DNV (2014) lists the fatality rate in road tunnels as 10^{-3} per tunnel year for Austria and France and 6.2×10^{-3} per tunnel year and vehicle kilometre for Germany. Schubert et al. (2011) give a fatality rate of approx. 10 per billion vehicle kilometres as an example. Diamantidis et al. (2000), Stille (2017), Zulauf (2012) and Shin et al. (2009) point out the limited applicability of classical target failure probabilities in tunnelling, see also Beard (2010).

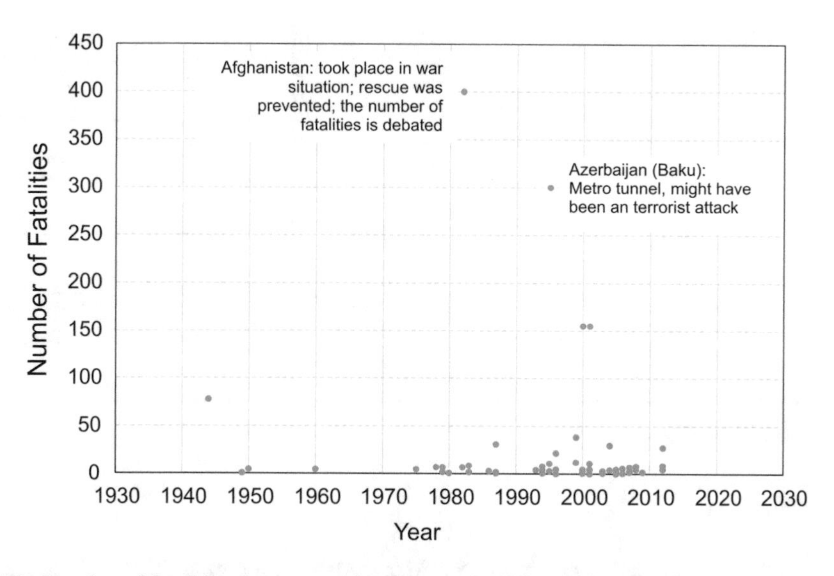

Fig. 5.8 Number of fatalities due to tunnel collapses (Spyridis and Proske 2021)

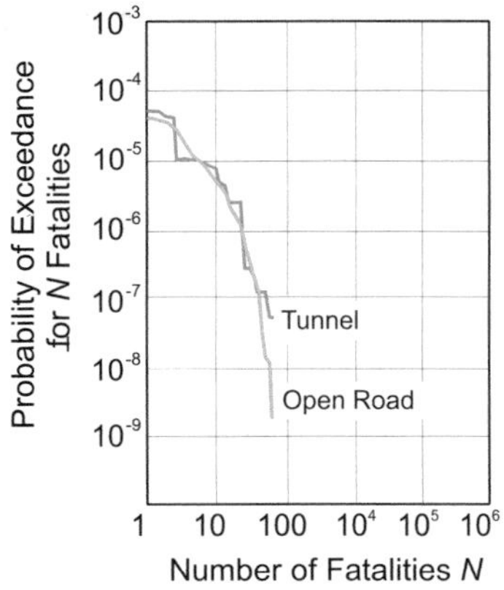

Fig. 5.9 Comparison of the distribution of fatality rates for open road and tunnel in an *F–N* diagram (Kröger and Høj 2000)

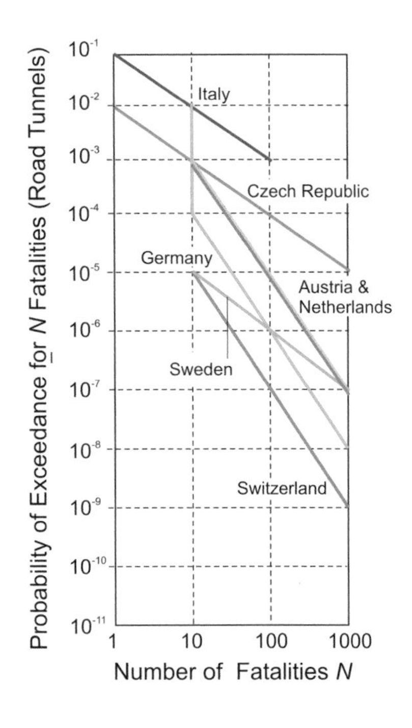

Fig. 5.10 Safety verification curves for road tunnels in different countries

5.9 Summary

In summary

- Due to the limited data available in the first study (Proske et al. 2019), the mean collapse frequencies and the dispersion could only be estimated reliably to a limited extent. By expanding the sample size, the quality of the results of the second study

(Spyridis and Proske 2021) with regard to mean collapse frequency and standard deviation is significantly better.

- The sample size of the failure probabilities was considerably expanded in the second study. The statistical moments of the failure probabilities should be significantly more credible in the second study.
- While in the first study the mean values of the collapse frequencies and the failure probabilities agreed very well, in the second study with the larger sample size a difference between calculation and observation became visible. However, the calculation results are on the safe side.
- The causes of collapse in terms of actions are known.
- Only a small proportion of tunnel collapses involve fatalities, as the majority of tunnel collapses occur during construction. However, due to the limited escape possibilities, accidents with high numbers of victims are known.
- This chapter uses results and data from Proske et al. (2019) and Spyridis and Proske (2021).

References

ABBV (2010) Ablösungsbeträge-Berechnungsverordnung, Bundesamt für Justiz. https://www.gesetze-im-internet.de/abbv/BJNR085600010.html. Accessed 13. April 2021

Athanasopoulou A, Bogusz W, Bournas D, Dimova S, Frank R, Sousa ML, Pinto A (eds) (2019) Standardisation needs for the design of underground structures. European Commission-JRC Technical Reports

Beard AN (2010) Tunnel safety, risk assessment and decision-making. Tunn Undergr Space Technol 25:91–94

Bergmeister K (2007) Transalpine tunnels: project overview and material management for the Brenner Base Tunnel. Beton- und Stahlbetonbau 1:19–23

Bergmeister K (2010) Brenner Base tunnel: current status. Tunnel 1:6–17

Bergmeister K (2016) Performance-based design of bridges and the brenner base tunnel. In: Beushausen H (ed) Performance-based approaches for concrete structures. fib symposium 2016, International Federation for Structural Concrete

Bjureland W, Spross J, Johansson F, Prästings A, Larsson S (2017) Reliability aspects of rock tunnel design with the observational method. Int J Rock Mech Min Sci 98:02–110

Breitenbücher R, Gehlen C, Schiessl P, Van Den Hoonaard J, Siemes T (1999) Service life design for the western scheldt tunnel. Durability of building materials and components. NRC Research Press, Ottawa, pp 2–15

BTS (2003) The british tunnelling society: the joint code of practice for risk management of tunnel works in the UK. BTS, London

CD 352 (2020) Design of road tunnels, Rev. 0. London

CEDD (2015) Catalogue of notable tunnel failure case histories (up to April 2015). Civil Engineering and Development Department, Hong Kong

Diamantidis D, Zuccerelli F, Westhäuser A (2000) Safety of long railway tunnels. Reliab Eng Syst Saf 67:135–145

DIN 1076 (1999) Engineering structures in the course of roads and paths—monitoring and testing, November 1999

DNV (2014) Risk level and acceptance criteria for passenger ships. First interim report, part 2: risk acceptance criteria European maritime safety agency, EMSA/OP/10/2013, Hovik

DoT (2020) National Tunnel Inventory. US Department of Transportation. https://www.fhwa.dot.gov/bridge/inspection/tunnel/inventory/download.cfm. Accessed 13. April 2021

EBP (2014) Review of the overall monitoring process, vulnerability analysis sub-project, risk analysis for the critical element "tunnel." Ernst Basler and Partner, Zollikon

FEDRO (2014) Guideline risk analysis for tunnels of national roads 19004, edition V1.10. Federal Roads Office, Switzerland

FEDRO (2016) Netzzustandsbericht der Nationalstrassen. Federal Roads Office, Switzerland

Fortsakis P, Litsas D, Kavvadas M, Trezos C (2011) Reliability analysis of tunnel final lining. Proceedings of the 3rd International Symposium on Geotechnical Safety and Risk, Munich

Fuyong C, Lin W, Wengang Z (2019) Reliability assessment on stability of tunnelling perpendicularly beneath an existing tunnel considering spatial variabilities of rock mass properties. Tunn Undergr Space Technol 88:276–289

Gharouni-Nik M, Naeimi M, Ahadi S, Alimoradi Z (2014) Reliability analysis of idealized tunnel support system using probability-based methods with case studies. Int J Adv Struct Eng 6(2):1

GlobalData (2019) Project Insight—Global Tunnels Construction Projects. GBDT16107300, October 2019, 53 pp. https://www.marketresearch.com/GlobalData-v3648/Project-Insight-Global-Tunnels-Construction-12849561/. Accessed 13. April 2021

Goh ATC, Hefney AM (2010) Reliability assessment of EPB tunnel-related settlement. Geomech Eng 2(1):57–69

Goh ATC, Zhang W (2012) Reliability assessment of stability of underground rock caverns. Int J Rock Mech Min Sci 55:57–163

Hamrouni A, Dias D, Sbartai B (2017) Reliability analysis of shallow tunnels using the response surface methodology. Undergr Space 2(4):246–258

Ingason H, Li YZ, Lönnermark A (2015) Tunnel fire dynamics. Springer, New York

ITA (2016) Tunnel market survey 2016. International tunneling and underground space association (ITA). https://www.tunnel-online.info/en/artikel/tunnel_Tunnel_Market_Survey_2016_3051818.html. Accessed 13. April 2021

ITA (2017a) Tunnelling market study 2016. Conferences, Tunnels, 7, International Tunneling and Underground Space Association, Lausanne

ITA (2017b) The world tunnel congress 2017. International Tunneling and Underground Space Association, Lausanne

ITA (2019) International tunneling and underground space association. Chatelaine, Switzerland. https://www.ita-aites.org/. Accessed 13. April 2021

Johansson F, Bjureland W, Spross J (2016) Application of reliability-based design methods to underground excavation in rock. Rock Engineering Research Foundation, Report No. 155, Stockholm

Kohno S, Ang AH, Tang WH (1992) Reliability evaluation of idealized tunnel systems. Struct Saf 11(2):81–93

Konstantis T, Konstantis S, Spyridis P (2016) Tunnel losses, causes, impact, trends and risk engineering management. ITA—AITES, WTC 2016, The World Tunnel Congress Including NAT2016, San Francisco, 22–28 Apr 2016

Kroetz HM, Do NA, Dias D, Beck AT (2018) Reliability of tunnel lining design using the hyperstatic reaction method. Tunn Undergr Space Technol 77:59–67

Kröger W, Høj NP (2000) Risk analyses of transportation on road and railway. Proceedings—part 2/2 of promotion of technical harmonization on risk-based decision-making, Workshop, Stresa, Italy, May 2000

Langford JC, Diederichs MS (2013) Reliability based approach to tunnel lining design using a modified point estimate method. Int J Rock Mech Min Sci 60:263–276

Laso E, Lera MSG, Alarcón E (1995) A level II reliability approach to tunnel support design. Appl Math Model 19(6):371–382

Li HZ, Low BK (2010) Reliability analysis of circular tunnel under hydrostatic stress field. Comput Geotech 37(1/2):50–58

Li X, Li X, Su Y (2016) A hybrid approach combining uniform design and support vector machine to probabilistic tunnel stability assessment. Struct Saf 61:22–42

Liu H, Low BK (2017) System reliability analysis of tunnels reinforced by rockbolts. Tunn Undergr Space Technol 65:155–166

Low BK, Einstein HH (2013) Reliability analysis of roof wedges and rockbolt forces in tunnels. Tunn Undergr Space Technol 38:1–10

Lü Q, Sun H-Y, Low B-K (2011) Reliability analysis of ground-support interaction in circular tunnels using the response surface method. Int J Rock Mech Min Sci 48:1329–1343

Lü Q, Chan CL, Low BK (2012) Probabilistic evaluation of ground-support interaction for deep rock excavation using artificial neural network and uniform design. Tunn Undergr Space Technol 32:1–18

Lü Q, Chan CL, Low BK (2013) System reliability assessment for a rock tunnel with multiple failure modes. Rock Mech Rock Eng 46(4):821–833

Lü Q, Xiao ZP, Ji J, Zheng J (2017) Reliability based design optimization for a rock tunnel support system with multiple failure modes using response surface method. Tunn Undergr Space Technol 70:1–10

Maidl B, Thewes M, Maidl U (2014) Handbook of tunnel engineering II. Basics and additional services for design and construction. Ernst & Sohn, Berlin

Meschke G, Cao B-T, Freitag S (2018) Reliability analysis and real-time predictions in mechanized tunneling. Resilience Engineering for Urban Tunnels. ASCE, Reston, pp 13–28

Miro S, König M, Hartmann D, Schanz T (2015) A probabilistic analysis of subsoil parameters uncertainty impacts on tunnel-induced ground movements with a back-analysis study. Comput Geotech 68:38–53

Mollon G, Dias D, Soubra AH (2009) Probabilistic analysis of circular tunnels in homogeneous soil using response surface methodology. J Geotech Geoenviron Eng 135(9):1314–1325

Papaioannou I, Heidkamp H, Düster A, Rank E, Katz C (2009) Random field reliability analysis as a means for risk assessment in tunnelling. In: Proceedings of 2nd international conference on computational methods in tunnelling EURO: TUN 2009

PR Newswire (2020) Tunnel and metro market 2020–2026 latest in-depth analysis report segment by manufacturers, type, applications and global forecast. https://www.prnewswire.com/in/news-releases/tunnel-and-metro-market-2020-2026-latest-in-depth-analysis-report-segment-by-manufacturers-type-applications-and-global-forecast-839650087.html. Accessed 8 Jan 2020

Proske D (2009) Catalogue of risks. Springer, Heidelberg

Proske D (2019) Comparison of frequencies and probabilities of failure in engineering sciences. In: Beer M, Zio E (eds) Proceedings of the 29th European Safety and Reliability Conference (ESRA), Hannover, pp 2040–2044

Proske D (2020) Erweiterter Vergleich der Versagenswahrscheinlichkeit und -häufigkeit von Kernkraftwerken, Brücken. Dämmen und Tunneln. Bauingenieur 95(9):308–317

Proske D, Spyridis P, Heinzelmann L (2019) Comparison of tunnel failure frequencies and failure probabilities. In: D Yurchenko, D Proske (eds) Proceedings of the 17th international probabilistic workshop, Heriot Watt University, Edinburgh, 11–13 Sept 2019, pp 177–182

Reiner H (2011) Developments in the tunnelling industry following introduction of the tunnelling code of practice, Presentation in IMIA annual conference, 21 September 2011. Munich Re, Amsterdam

Sandström GE (1963) The history of tunneling: underground working through the ages. Barrie and Rockliff, London

SBB (2018) Die SBB in Zahlen und Fakten 2017. Schweizer Bundesbahn, Switzerland

Schäfer M (2019) Tunnelling in Germany: statistics (2018/2019). Tunnel 6:8–19. https://www.stuva.de/?id=statistik. Accessed 13. April 2021

Schubert M, Faber MH (2008) Assessment of risks and criteria for defining accepted risks as a result of exceptional impacts in engineering structures. FEDRO, Federal Roads Office, Berne

Schubert M, Høj NP, Köhler J, Faber MH (2011) Development of a best practice methodology for risk assessment in road tunnels. ASTRA, Federal Roads Office (FEDRO), Bern

Seidenfuss T (2006) Collapse in tunnelling. MSc Degree Thesis, Stuttgart University of Applied Sciences and EPFL Lausanne, Lausanne

Shin H-S, Kwon Y-C, Jung Y-S, Bae G-J, Kim Y-G (2009) Methodology for quantitative hazard assessment for tunnel collapses based on case histories in Korea. Int J Rock Mech Min Sci 46:1072–1087

SIA 198 (2004) Untertagbau—Ausführung. Schweizerischer Ingenieur- und Architektenverein, Zürich

Sousa RL (2010) Risk analysis for tunneling projects. Doctoral thesis, MIT

Špačková O, Šejnoha J, Straub D (2013) Probabilistic assessment of tunnel construction performance based on data. Tunn Undergr Space Technol 37:62–78

Spyridis P (2014) Adjustment of tunnel lining service life through appropriate safety factors. Tunn Undergr Space Technol 40:324–332

Spyridis P, Konstantis S, Gakis A (2016) Performance indicator of tunnel linings under geotechnical uncertainty. Geomech Tunn. 9(2):158–164

Spyridis P, Proske D (2021) Revised comparison of tunnel collapse frequencies and tunnel failure probabilities. ASCE-ASME J Risk Uncertainty Eng Syst, Part A: Civ Eng 7(2):04021004-1–04021004-9

Statista (2020) Number of road tunnels in China from 2010 to 2018, Length of road tunnels in Japan from 2010 to 2017 (in kilometres), Number of Deutsche Bahn AG tunnels in Germany from 2012 to 2019. https://de.statista.com/. Accessed 13. April 2021

Stille H (2017) Geological uncertainties in tunnelling—risk assessment and quality assurance. Sir Muir Wood Lecture 2017, Lausanne, Apr 2017

Striegler W (1993). Tunnelbau s. l. Verlag für Bauwesen, Berlin

STS (2020) Anzahl der Tunnel und Stollen. Swiss tunneling society. https://www.swisstunnel.ch/tunnelbau-schweiz/uebersichtsgrafiken/number-tunnel/. Accessed 13. April 2021

Su YH, Zhang P, Zhao MH (2007) Improved response surface method and its application in stability reliability degree analysis of tunnel surrounding rock. J Cent South Univ Tech 14(6):870–876

Thewes M, Maidl U (2013) Handbook of tunnel engineering: volume I. Structures and methods. Ernst & Sohn, Berlin

Vogel T, Zwicky D, Joray D, Diggelmann M, Høj NP (2009) Tragsicherheit der bestehenden Kunstbauten, Sicherheit des Verkehrsssystems Strasse und dessen Kunstbauten. Federal Roads Office, Bern, Dec 2009

Wang Q, Fang H, Shen L (2016) Reliability analysis of tunnels using a metamodeling technique based on augmented radial basis functions. Tunn Undergr Space Technol 56:45–53

Wikipedia (2020) List of countries by total road tunnel length. https://en.wikipedia.org/wiki/List_of_countries_by_total_road_tunnel_length.Accessed 13. April 2021

Yang W, Baji H, Li CQ (2018) Time-dependent reliability method for service life prediction of reinforced concrete shield metro tunnels. Struct Infrastruct Eng 14(8):1095–1107

Yang XL, Zhou T, Li WT (2018) Reliability analysis of tunnel roof in layered Hoek-Brown rock masses. Comput Geotech 104:302–309

Zeng P, Senent S, Jimenez R (2016) Reliability analysis of circular tunnel face stability obeying Hoek-Brown failure criterion considering different distribution types and correlation structures. J Comput Civ Eng 30(1):04014126

Zhang G-H, Jiao Y-Y, Chen L-B, Wang H, Li S-C (2016) Analytical model for assessing collapse risk during mountain tunnel construction. Can Geotech J 53:326–342

Zhao H, Ru Z, Chang X, Yin S, Li S (2014) Reliability analysis of tunnel using least square support vector machine. Tunn Undergr Space Technol 41:14–23

Zhao L (2009) Statistics of tunnel collapse accidents and its assessment based on case reasoning. J Railw Sci Eng 6(4):54–58

Zulauf C (2012) Risk evaluation for road tunnels: current developments. Proceedings of the 6th international conference tunnel safety and ventilation 2012, Graz, pp 26–33

Retaining Structures

6

6.1 Introduction

Like bridges, tunnels, or dams, retaining structures belong to the infrastructure. Retaining structures were certainly used during or shortly after the introduction of stone buildings. For this reason, it is estimated here that the first retaining structures were built at least about 6,000 years ago.

6.2 Definition of Retaining Structures

Four definitions of retaining structures are given below. Sometimes the term supporting structure is used.

Reban (2015) defines retaining structures as follows: "*Retaining structures include all types of walls or support systems where structural components are stressed by forces from the supported material*".

The definition according to FHWA (1997) is "*The purpose of a retaining structure is to stabilise an otherwise unstable soil mass by providing lateral support or reinforcement*".

Deutsche Bahn (DB 2015) defines "*Retaining structures are supporting structures that take horizontal and vertical loads (dynamic and/or static) from the ground and transfer them to the ground at the base*".

DIN 1076 (1999) defines retaining structures as "*engineering structures that perform a supporting function with respect to the ground, the road body or bodies of water and have a visible height of 1.5 m or more*".

D. Proske, *The Collapse Frequency of Structures*,
https://doi.org/10.1007/978-3-030-97247-9_6

6.3 Stock of Retaining Structures

According to Proske (2020), there are at least 26 million retaining structures worldwide. Table 6.1 shows the number and partly the length of retaining structures in different industrialised countries.

6.4 Calculation of Collapse Frequencies

6.4.1 Introduction

Various individual case studies of retaining structures exist, but summary reports were only a few available. The following references were used Kim (1995), ARUP (2002), Cheng and Ko (2010), Lo and Cheung (2014). In addition, seven own values were calculated with the data from Schneider et al. (2014) for earthquakes.

Table 6.1 Number of retaining structures in different countries and for different infrastructure managers (Hofmann et al. 2021)

Region or country	Infrastructure manager	Number	Kilometres	Reference
European Union			50,000	EC (2004)
Germany	German railways	10,033	1,229	DB AG (2017)
Austria	Total	140,000		Nöhrer (2019)
Austria	ÖBB	10,000		Reban (2015), ASFINAG (2020)
Austria	ASFINAG	961		ASFINAG (2020)
Switzerland	Total	50,000		Proske (2020)
Switzerland	SBB	11,000	413	SBB (2020), Friedl (2019)
Switzerland	FEDRO	2,500		Loeb (2020)
Japan	Railway	7,989		Shinoda et al. (2013)
Czech Republic		5,543		EC (2004)
Denmark	National roads	18		EC (2004)
France	National roads	13,729		EC (2004)
Spain	National roads	3,641		EC (2004)
Sweden	National roads and streets in Stockholm	600		EC (2004)
United Kingdom			4,433	EC (2004)

6.4.2 Central Estimator

Figure 6.1 shows the determined collapse frequencies and calculated failure probabilities of retaining structures. Compared to other structures, retaining structures show a high average collapse frequency.

6.4.3 Measure of Deviation

Neither a standard deviation nor a confidence interval for the mean values for retaining structures has been calculated so far.

6.4.4 Trend

In contrast to the other types of structures, such as bridges or dams, no trend could be calculated here because the scatter of data points increases with decreasing time interval. In other words, the scatter of data points from 2005 to 2015 masks a possible trend. The case is roughly comparable with the evaluations of the tunnel structures.

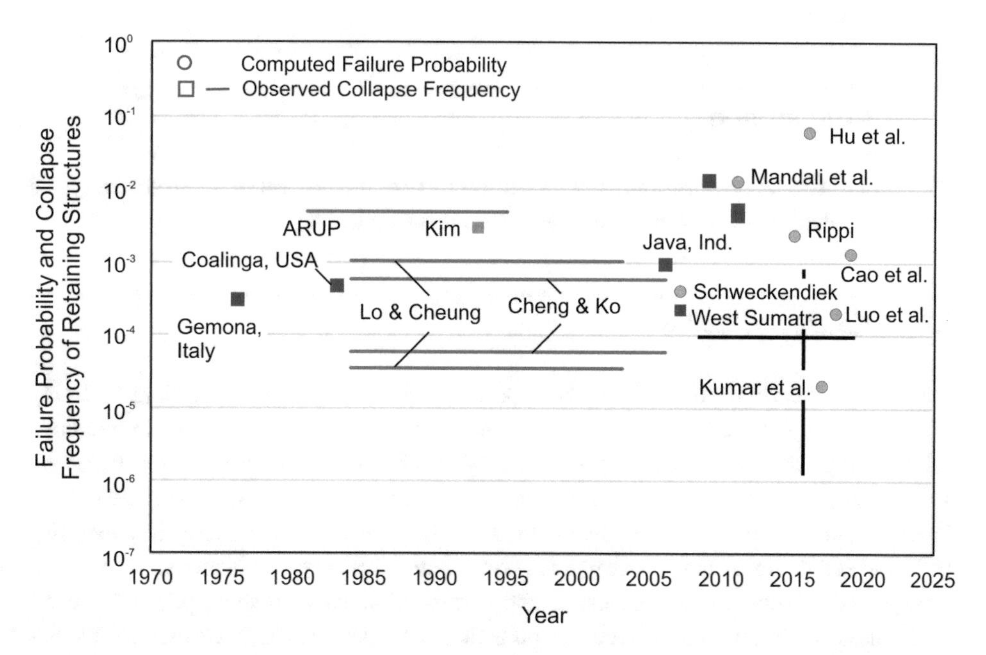

Fig. 6.1 Collapse frequencies and failure probabilities of retaining structures (Hofmann et al. 2021)

Table 6.2 Mean values (Hofmann et al. 2021)

Probabilistic calculations	Observed collapse frequencies	Target value per year
1.35×10^{-3}	8.14×10^{-4}	10^{-6}

Table 6.3 Classification of retaining structures based on Boley (2012)

Examples of retaining walls with shallow foundations	Examples of retaining walls with deep foundations
Gravity retaining wall	Sheet pile wall
Cantilever retaining wall	Girder pile wall
Gabion retaining walls	Slurry wall
Anchor wall	Pile wall
Chemise wall	Bar walls
Reinforce soil	Needling and grouting

6.5 Calculation of Failure Probabilities

Seven publications were used from which the probability of failure was taken (Schweckendiek et al. 2007; Mandali et al. 2011; Rippi 2015; Hu et al. 2016; Kumar and Roy 2017; Luo 2018; Cao et al. 2019).

6.6 Comparison

The mean observed collapse frequencies during use are the highest of all structure types. The mean calculated failure probabilities reflect this (Table 6.2).

6.7 Causes of Collapse

The causes of collapse depend on the type of retaining structure. According to Boley (2012), retaining structures can be divided into retaining walls with shallow foundations and retaining walls with deep foundations. Table 6.3 shows the classification based on Boley (2012). Schneider et al. (2014) and Adam et al. (2019) show further classifications of retaining structures. The Swiss Railway (SBB), for example, recognises massive, wall-like, and flexible retaining structures and constructive slope protections.

Figure 6.2 shows a classification of the causes of damage to sheet piles. However, this focuses solely on human error, not on actions or other causes. In contrast, Table 6.4 shows different decisive factors for collapses of retaining structures from three studies.

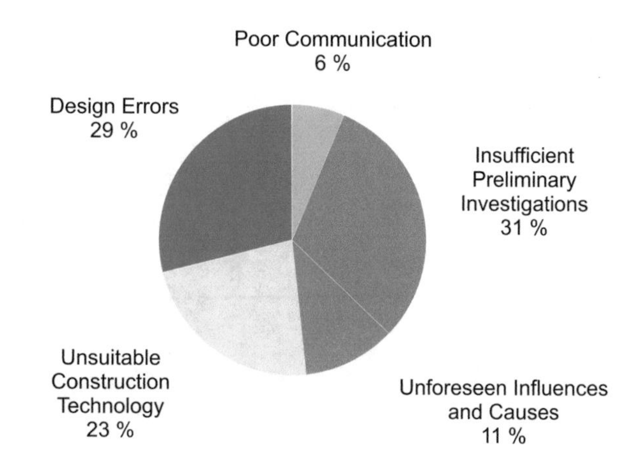

Fig. 6.2 Percentage distribution of the causes of structural damage using the example of steel sheet piling (Rizkallah et al. 1990)

Table 6.4 Dependence of collapses on the different decisive factors in different models

Parameter	Study I	Study II	Study III
Condition	×		×
Wall type	×		×
Material	×		
Wall height	×	×	×[1]
Inclination	×	×	
Precipitation amount	×		[2]
Geological zone	×		
Uphill/downhill	×		
Other			×[3]

[1] Anchorage.
[2] No, but the proximity of water is taken into account.
[3] Preload and monitorability.

6.8 Mortality and Casualty Figures

Although retaining structures seem to fail more frequently than other types of structures, the consequences seem to be smaller for various reasons. Based on data from ARUP (2002) and Curbach and Proske (2004), Fig. 6.3 shows two *F-N* curves, one for retaining structures and one for bridges. It can be seen that the curve for the retaining structures starts earlier but then falls faster than the curve for bridges. *F-N* curves are a common representation of risks to people and goods (Proske 2022).

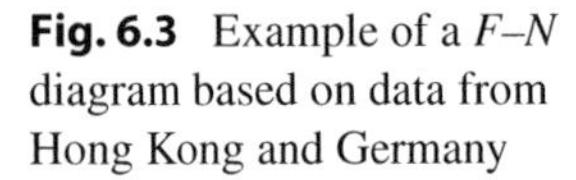
Fig. 6.3 Example of a *F–N* diagram based on data from Hong Kong and Germany

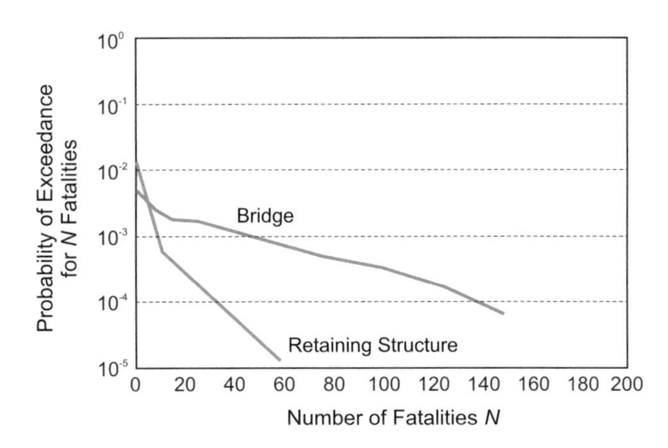

6.9 Summary

Retaining structures fail significantly more often than other types of structures, but the impact in terms of fatalities seems to be lower than for other types of structures. This can be due to various reasons, such as the good visibility of material on the roadway or the loss of the roadway. Nevertheless, there are known cases where the collapse of retaining structures has caused fatalities, such as the tragic fatality on the Brenner motorway in 2012. However, retaining structures can also be destroyed by natural hazard processes, such as landslides, mudflows, or earthquakes. Then larger numbers of victims are possible.

Compared to the bridges, tunnels and dams, the causes of collapse are less consistently named and less clearly elaborated.

References

Adam D, Bergmeister K, Florineth F (2019) Stützbauwerke, Betonkalender 2019: Parkbauten, Geotechnik und Eurocode 7. In: Bergmeister K, Fingerloos F (eds) Ernst und Sohn, vol 108. Berlin, pp 367–454

ARUP (2002) QRA of collapses and excessive displacement of deep excavations. Ove Arup and Partners Hong Kong Limited, Geotechnical Engineering Office, GEO Report No. 124, Feb 2002

ASFINAG (2020) Research project SIBS: Safety assessment of existing retaining structures (Unpublished)

Boley C (2012) Handbuch Geotechnik. Vieweg + Teubner Verlag, Wiesbaden, p 2012

Cao ZJ, Gao GH, Li DQ, Wang Y (2019) Values of Monte Carlo samples for geotechnical reliability-based design, Conference: Proceedings of the 7th international symposium on geotechnical safety and risk (ISGSR 2019) Vol 19, Dec 2019, pp 86–95

Cheng PFK, Ko FWY (2010) An updated assessment of landslide risk posed by man-made slopes and natural hillslides in Hong Kong, Geotechnical Engineering Office, GEO Report No. 252, July 2010
Curbach M, Proske D (2004) Risk investigation using the example of historical bridges under ship impact. Beton- und Stahlbetonbau 99(12):956–966
DB (2015) Planning, execution and maintenance of retaining structures on structural facilities of DB Station and Service AG, Technical Note TM 2015–03 I.SBB, 10.6.2015
DB AG (2017) Performance and financing agreement: infrastructure condition and development report 2016, Deutsche Bahn—Communication Infrastructure, Berlin, Sept 2017
DIN 1076 (1999) Engineering structures in the course of roads and paths—Monitoring and testing, November 1999
EC (2004) COST 345: Procedures required for the assessment of highway structures working group 1 report, European Commission, Directorate General Transport and Energy, Report, 2004
FHWA (1997) Geotechnical Engineering Circular No. 2—Earth Retaining Systems, U.S. Department of Transportation—Federal Highway Administration, Technical Report, Feb 1997
Friedl H (2019) Challenges and directions in asset management for SBB bridges. Burgdorfer Brückenbautag 2019, Bern University of applied sciences and construction and knowledge, conference proceedings, Sept 2019
Hofmann C, Proske D, Zeck K (2021) Vergleich der Einsturzhäufigkeit und Versagenswahrscheinlichkeit von Stützbauwerken. Bautechnik 98(7):475–481.
Hu QF, Zhang DM, Huang HW, Phoon KK (2016) Fully probabilistic-based design of cantilever wall in spatially varied clay. In: 6th Asian-Pacific symposium on structural reliability and its application (APSSRA 2016), pp 433–438
Kim J (1995) Reliability-based design of a retaining wall. PhD thesis, University of Virginia, USA
Kumar A, Roy P (2017) Reliability analysis of retaining wall using imprecise probability, safety, reliability, risk, resilience and sustainability of structures and infrastructure. In: Bucher C, Ellingwood BR, Frangopol DM (eds) 12th International conference on structural safety and reliability, Vienna, Austria, 6–10 Aug 2017 TU-Verlag Vienna, pp 288–297
Lo D, Cheung W (2014) Assessment of landslide risk of man-made slopes in Hong Kong, Geotechnical Engineering Office, GEO Report No. 177, Hong Kong, July 2014
Loeb R (2020) 500 critical retaining structures must be strengthened. VSS Online, pp 5. https://www.vss.ch/fileadmin/redacteur/e-paper_SuV_11_17
Luo Z, Hu B (2018) Life-cycle reliability-based assessment of internal stability for mechanically stabilized earth walls in a heavy haul railway. Comput Geotech 10:141–148
Mandali AK, Sujith MS, Rao BN, Maganti J (2011) Reliability analysis of counterfort retaining walls. Electron J Struct Eng 11(1):42–56
Nöhrer F, Marte R (2019) Risk management of retaining structures in the railway network of the province of Styria, Geomech Tunneling 12:515–522
Proske D (2020) The global health burden of structural failure. Civ Eng 97(4):233–242
Proske D (2022) Katalog der Risiken, 2. vollständig überarbeite Aufl. Springer, Berlin - Heidelberg
Reban M (2015) Ist-Zustandserfassung und Bewertung bestehender, unverankerter Stützbauwerke. Diploma thesis, Institute of Soil Mechanics and Foundation Engineering, TU Graz, Austria
Rippi A (2015) Structural reliability analysis of a dike with a sheet pile wall—coupling reliability methods with finite elements. M.S. thesis, TU Delft
Rizkallah V, Harder H, Jebe P, Vogel J (1990) Bauschäden im Spezialtiefbau, Institut für Bauschadensforschung e.V., Heft 3, Eigenverlag, Hannover
SBB (2020) Network Status Report 2019, SBB Infrastructure, March 2020
Schneider H, Quinteros S, Romer B, Seifert J (2014) Design and verification of retaining structures under earthquake action. Hochschule für Technik Rapperswil, April 2014

Schweckendiek T, Courage WMG, Van Gelder PHAJM (2007) Reliability of sheet pile walls and influence of corrosion—structural reliability analysis with finite elements. In: Aven T, Vinnem JE (eds) Risk, reliability and societal safety. Taylor and Francis Group, London, pp 1791–1799

Shinoda M, Najajima S, Abe K, Ehara T, Kubota Y (2013) Stability inspection method for existing retaining walls. Quarterly Report of RTRI 54(3):159–165

7 Buildings and Structures

7.1 Introduction

The first structures in the form of huts were probably erected at the beginning of the Neolithic period (Mann 1991). The first massive temple complex in Turkey was built more than 11,000 years ago (Mann 2011). Massive buildings are likely to have been erected regularly from around 9,000 years ago (Mann 1991; Bilham 2009) and to have formed the first settlements relatively quickly.

Today, buildings in Germany are worth over 14 trillion euros (ZIA 2017). This corresponds to 80% of Germany's total tangible assets and shows the great importance of buildings as a storage of wealth. In Switzerland, the replacement value of buildings is around 1.55 trillion Swiss francs (HEV 2016).

Structures and buildings are not only among the earliest and most capital-intensive, but also among the most numerous (see Sect. 7.3) and longest-lasting technical products. Structures such as the pyramids in Egypt with a lifespan of over 4,000 years provide impressive proof of this.

The right to shelter and thus the use of buildings is now even a human right according to the UN (United Nations 2018).

7.2 Definition of Buildings and Structures

In the following, three definitions are given for the terms "building structures", "buildings" and "buildings above ground level", which have the same meaning in the context of this chapter.

The Federal Statistical Office (Statistisches Bundesamt 2014) defines buildings as: "...*structures that generally rise substantially above the earth's surface. For technical*

D. Proske, *The Collapse Frequency of Structures*,
https://doi.org/10.1007/978-3-030-97247-9_7

reasons, building structures also include such independently usable underground structures that can be entered by people and are suitable or intended to serve the protection of people, animals or property (e.g……, underground shop centres and production facilities, underground car parks). "

The Swiss Confederation (Schweizerische Eidgenossenschaft 2017) defines in the Ordinance on the Federal Register of Buildings and Dwellings, Art. 2, paragraph b: "*Buildings [are] a permanent structure, provided with a roof, firmly attached to the ground, capable of accommodating persons and serving residential purposes or purposes of work, education, culture, sport or any other human activity; …*".

The Free State of Saxony (Freistaat Sachsen 2018) defines in the Saxon Building Code in § 2, paragraph 2: "*Buildings are independently usable, covered structural facilities that can be entered by people and are suitable or intended to serve the protection of people, animals or property.*"

7.3 Stock of Buildings and Structures

Data on building stock are often not uniform or only include subsets. Table 7.1 lists the building stock and other relevant parameters for determining the building stock for different countries. Table 7.2 presents a summary and extrapolation using the numbers from Table 7.1.

Different numbers in various references may be due to actual changes in the building stock or due to different counting. For Switzerland, for example, figures can be found in HEV (2016) and Baldegger et al. (2020) or the subset of 1.74 million residential buildings in Federal Statistical Office (Bundesamt für Statistik 2019).

Another example is the calculation of the building stock for Germany, which is based on the figures of 21.05 million residential buildings and 3.52 million non-residential buildings. This results in 24.57 million buildings (Schiller et al. 2015). However, in Destatis (2019) 18.9 million residential buildings and in Behnisch et al. (2012) 49 million geometry objects are mentioned, of which 2/3 are main buildings and 1/3 are secondary buildings, which corresponds to 32 million building ensembles and 22 million postal assignable addresses.

The determination of the number of buildings in Europe was based on the 117 million residential buildings in the EU according to Fabbri (2020) and the division of the building stock in the EU into 75% residential and 25% non-residential buildings (BPIE 2011).

The stock of buildings for China was back-calculated based on the available building areas according to Hong et al. (2016). A characteristic building area was estimated and converted into a number of buildings. Half of the German residential and office space per inhabitant was used as the characteristic building area.

So far, three values have been identified for the global stock of buildings: 1.1 billion in Proske & Schmid (2021), 1.3 billion in Proske (2020), 1.5 billion in Proske (2021).

Table 7.1 Number and area of buildings for different countries (Proske & Schmid 2021)

Land	Building Area in Billion m^2	Buildings in Million	Population in Million (2019/20) (DSW 2019)	Buildings per Inhabitants	Building Area per Inhabitants
Switzerland	0.6 (BPIE 2011)	2.75 (Baldegger et al. 2020)	8.6	0.32	69.8
Italy	2.95 (BPIE 2011)	43.19 (Agenzia nationale 2016)	60.3	0.72	48.1
Germany	4.3 (BPIE 2011)	21.7 (Dena 2019) -24.5 (Schiller et al. 2015)	83.1	0.26	51.8
Austria	0.4 (BPIE 2011)	2.19 (Statistik Austria 2015)	8.9	0.25	44.9
Czech Republic	0.39 (BPIE 2011)	2.41 (GTAI 2020)	10.7	0.23	37.4
US		158.54 (Statista 2020; CBECS 2015)	329	0.48	
Canada		13.3 (Openstreetmap 2019. Government of Canada 2020)	38.03	0.35	
Australia		15.2 (Malo 2018)	25.72	0.59	
Russia	5.4 (Sirviö & Illikainen 2015)		146.8		36.8
China	56.1 (Hong et al. 2016)	264 (Hong et al. 2016)	1,400.5	0.18	41.0
Japan		34 Tanikawa et al. 2015)	126.5	0.27	
Europe		146.3 (Fabbri 2020; BPIE 2011)	746	0.19	48.8 (BPIE 2011)
India		361 (Kaul 2015)	1,366	0.33	
Ghana		4.65 (GGS 2013)	30.7	0.15	

Table 7.2 Number and area of buildings for different continents (Proske & Schmid 2021)

	Region	Building Area in Billion m^2	Population in Million (2019) (DSW 2019)	Building Area per Inhabitants	Buildings in Million	Buildings per Inhabitants
Industrialized Countries	Europe	33.00	746	44.24	203.38	0.27
	North America	35.00	367	95.37	158.54	0.43
	Australia	2.84	25.3	112.22	15.24	0.60
	Total	70.84	1,138.3	251.83	377.16	
	Mean Value. weighted	32.97		62.23		0.33
Developing Countries	Asien	100.20	4,603.7	21.77	533.48	0.12
	Africa	24.00	1,305	18.39	127.78	0.10
	Latin Amerika	20.00	645	31.01	106.48	0.17
	Total	144.20	6,553.7	71.17	767.74	0.39
	Mean Value. weighted	22.68		22.56		0.12
	Worldwide		7,692		1,144.89	

However, the real number of buildings may differ from this because the definition of a building differs in different countries. For example, according to the UN (United Nations 2018), approximately one billion people live in slums. According to the Indian National Building Organisation (2016), over 100 million people live in 17.35 million buildings in Indian slums. Presumably, such buildings would not be considered buildings from a Western perspective.

Besides counting buildings, one can also try to determine the number of buildings via other parameters. One possibility is to calculate back using the gross domestic product per inhabitant, the living space or usable space per inhabitant or the mass. In fact, the total mass of all buildings on earth, including their technical equipment, is known. According to Krausmann et al. (2017), this value is 800 peta-gram.

The global usable area of buildings is also known. In 2017, it was 162.8 billion m^2 (Naviant Research 2018). This value is expected to increase to 183.5 billion m^2 by 2026 (Naviant Research 2018). Table 7.2 estimated a much larger total floor space of 215.04 billion m^2.

The high dynamics (see also Ernst 2006; Griess 2015) reflected in these figures is also reflected in the number of housing units. This doubled from 1950 to 2003 (Bilham 2009). The next doubling is expected by 2030. So, the doubling no longer takes 50 years, but only 30 years. Bilham (2009) estimated that there will be half a billion additional house units by 2050. Then the number of buildings worldwide would probably be 2 billion.

7.4 Calculation of Collapse Frequencies

7.4.1 Introduction

Building collapses are regularly reported. Usually after heavy storms, snowfalls, or earthquakes. However, collapses without such extreme natural events have occurred in recent decades, not only in developing countries, e.g. in Mumbai (India) in 2017, in Accra (Ghana) in 2012 and in Savar Upazila (Bangladesh) in 2013 with over 1,000 fatalities, but also in developed countries such as in Halstenbeck (Germany) in 1997 and 1998, in Bad Reichenhall (Germany) in 2006, in Cologne (Germany) in 2009, in Marseille (France) in 2018, in Naples (Italy) in 2017 or the collapse of the World Trade Center in New York (USA) in 2001 after a terrorist attack with over 3,000 fatalities.

Systematic studies have been produced for various countries, such as Denmark (Aagaard & Pedersen 2013), Switzerland (Ortega 2000), Canada (Allen & Schriever 1972), the USA (Wardhana & Hadipriono 2003; Eldukair & Ayyub 1991; Geis 2012), Nigeria (Hamma-adama & Kouider 2017, Omenihu et al. 2016, Ayedun et al. 2011 and Ghana and Kenya (Asante & Sasu 2018), South Africa (Emuze et al. 2015), Malaysia (Sulaima et al. 2014; Azhahar et al. 2011), Singapore and Thailand (Michael & Razak 2013) and internationally (Scheer 2001).

Figure 7.1 was constructed using numerous references. Some of the data points include own calculations using data from the 28 references (Lyu et al. 2018, Feifei 2014, National Crime Records Bureau 2020, Fagbenle & Oluwunmi 2010, Omenihu et al. 2016, Ayodeji 2011, Ede et al. 2018, Windapo & Rotimi 2012, Hamma-adama & Kouider 2017, Eldukair & Ayyub 1991, Wardhana & Hadipriono 2003, Hadipriono 1985, Swissinfo 2004, Tagesanzeiger 2011, Boateng 2012, Luzerner Zeitung 2019, Asante & Sasu 2018, meinbezirk 2012, ORF Tirol 2015, Feuerwehr Trieben 2013, Feuerwehr Wien 2020, Tageblatt Frauenfeld 2017, Stuttgarter Nachrichten 2019, Liebers 1996, RP Online I 2020, RP Online II 2005, Baunetz.de 1998, Zerna 1983).

Figure 7.2 presents a summary of the individual data from Fig. 7.1. Even though the division into industrialised and developing countries is being used less, it has been used here as a simplification. It may also be critically questioned whether China can still be considered a developing country. China is already the world's largest economy in purchasing power terms, and in a few years, it will also be the world's largest economy in absolute terms.

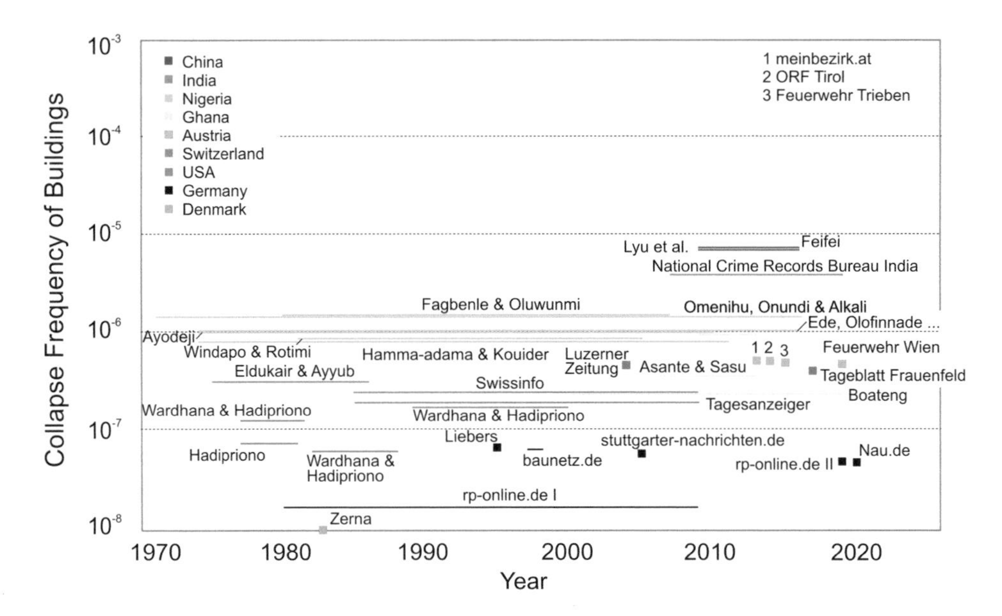

Fig. 7.1 Collapse frequencies of buildings with all references (Proske & Schmid 2021)

However, it is important to note that industrialised and developing countries differ not only in terms of income, technology standards and productivity, but also in terms of qualification and training levels (Duden 2016).

7.4.2 Central Estimator

Based on the available data, a mean value of the collapse frequency in industrialised countries of 2.4×10^{-7} and of 4.7×10^{-6} in developing countries was determined. The global weighted average is 3.3×10^{-6}.

7.4.3 Measure of Deviation

The calculation of the coefficient of variation of the values in Fig. 7.1 is about 1.27. The range of the data is considerable and covers almost three orders of magnitude, from about 10^{-5} per year to 10^{-8} per year. Figure 7.3 shows the histogram of the observed collapse probabilities and the calculated failure probabilities. However, the evaluation must be critically questioned. It is surprising, for example, that the collapse frequency in Fig. 7.2 is practically identical in Ghana and Austria, but the collapse frequency in

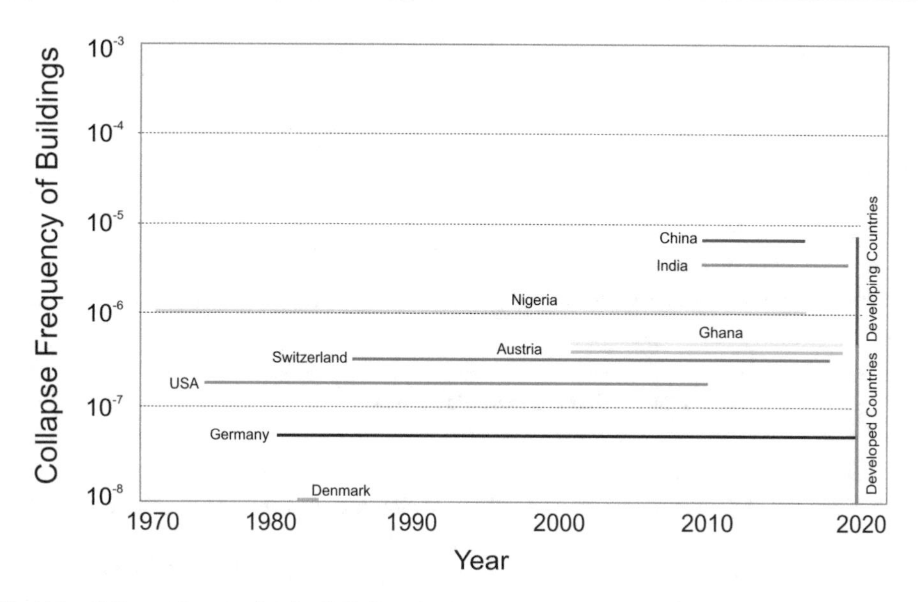

Fig. 7.2 Collapse frequencies for buildings for different countries (Proske & Schmid 2021)

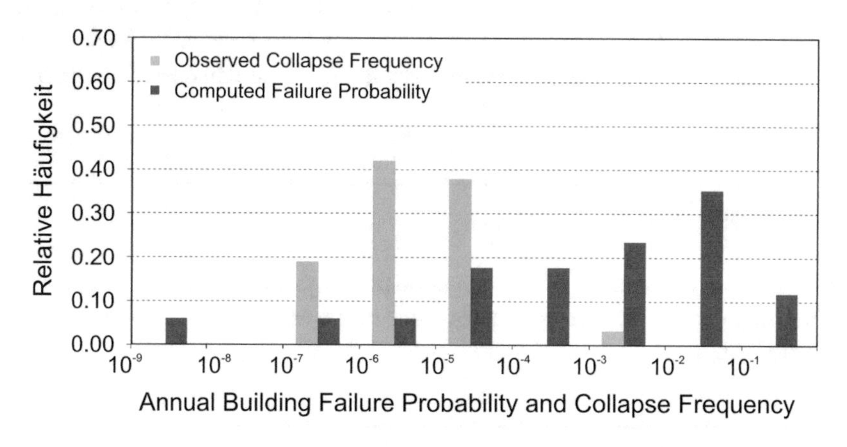

Fig. 7.3 Histogram of the calculated failure probabilities and the observed collapse frequencies of buildings

Denmark and China differs by a factor of 1,000. Further investigations, e.g. on underreporting, are necessary here.

The very high observed collapse frequency is due to large-scale accidental actions, such as earthquakes or floods.

7.4.4 Trend

Based on the data over a period of 50 years, no trend can be observed. This can have various causes. For example, the time series may be too short to detect changes.

In addition, the considerable construction activities in developing countries and especially in Africa and Asia may distort the trend (Ernst 2006 and Griess 2015), because just as with bridges and tunnels, a significant proportion of collapses occur during and shortly after the construction phase.

7.5 Calculation of Failure Probabilities

An evaluation of probabilistic calculations for buildings and structures has been carried out but has not yet been published (Proske and Schmid 2022). Overall, the calculations show an extraordinarily large scatter, just as with the tunnels.

The values range from 1.5×10^{-19} to 2×10^{-1} per year (Fig. 7.4). The mean value is 7.9×10^{-3} and the median 1.06×10^{-5} for all calculations without accidental loads. The scatter is mainly due to the probabilistic calculations themselves, because considering accidental actions such as earthquakes or hurricanes leads to a reduction in the scatter. Thus, the mean and median for the results of all probabilistic calculations are 8.4×10^{-3} and 6×10^{-4}, respectively. 22 probabilistic calculations were considered. Further details can be found in Proske and Schmid (2022).

7.6 Comparison

Table 7.3 shows a comparison of the probabilistic calculations and the collapse frequencies. The comparison shows that the mean value of the calculations is above the observations (Table 7.3).

7.7 Causes of Collapse

The causes of structural collapses are manifold and can for example be divided into the type of human error and the type of the action. Figs. 7.5 and 7.6 list the causes of building collapses without collapses due to large-scale accidental actions.

When comparing the figures, the effects of different qualification and training levels of those involved in construction in industrialised and developing countries become clearly visible. This is particularly worrying in view of the extraordinarily high level of construction activity in developing countries (Ernst 2006 and Griess 2015). In addition, modern building standards from industrialised countries are often adopted in developing

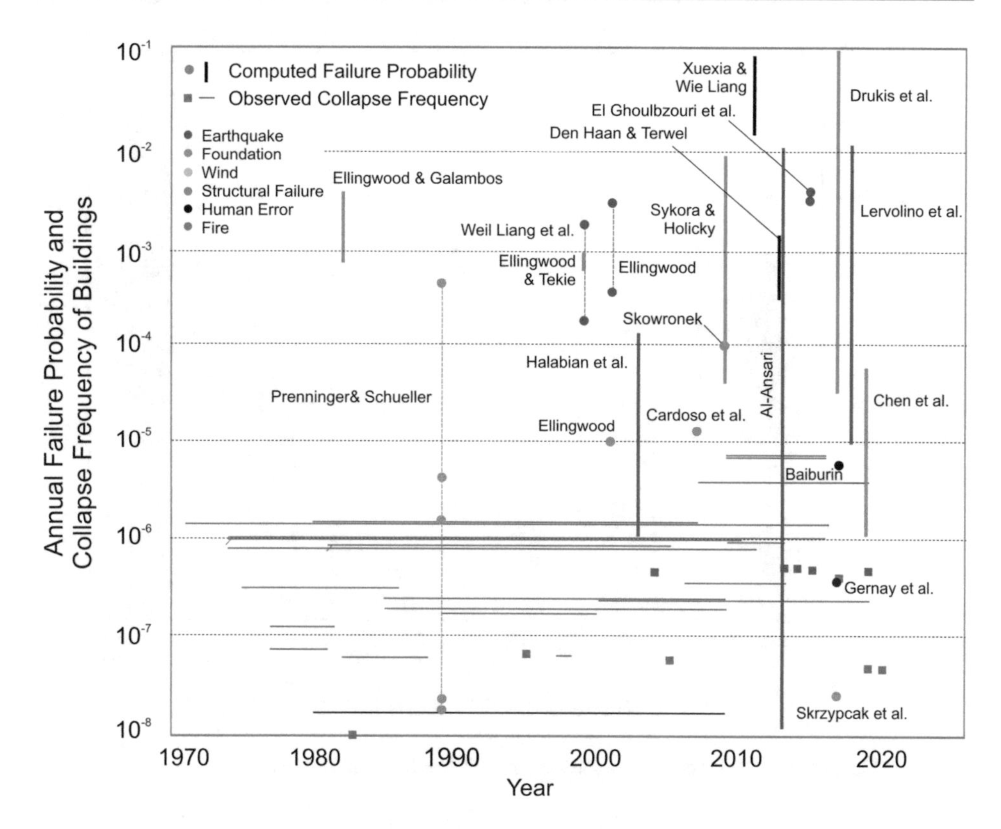

Fig. 7.4 Probabilities of failure and collapse frequencies for buildings according to Proske and Schmid (2022). Collapse frequencies are in grey colour and without references to improve the readability

Table 7.3 Mean values of the probabilistic calculations and the collapse frequencies (according to Proske & Schmid 2021 and Proske & Schmid 2022)

Probabilistic calculations	Observed frequency
1.06×10^{-5} $(1.5 \times 10^{-19} \ldots$ $1 \times 10^{-1})$	3.30×10^{-6} $(2.7 \times 10^{-7}$ $4.7 \times 10^{-6})$

countries. However, modern building standards categorically exclude human error, whether through the requirements formulated in the standards for the qualifications of those involved or through organisational measures such as the four-eyes principle for load-bearing capacity verifications. This assumption is based on Fig. 7.5 not fulfilled in developing countries.

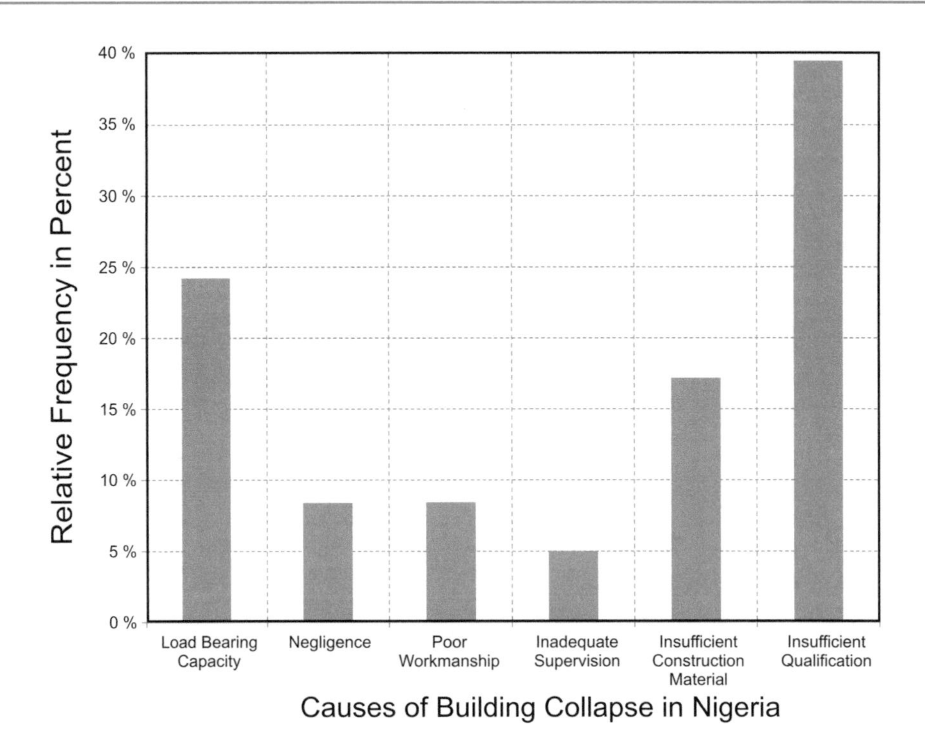

Fig. 7.5 Causes of building collapses according to Omenihu et al. (2016)

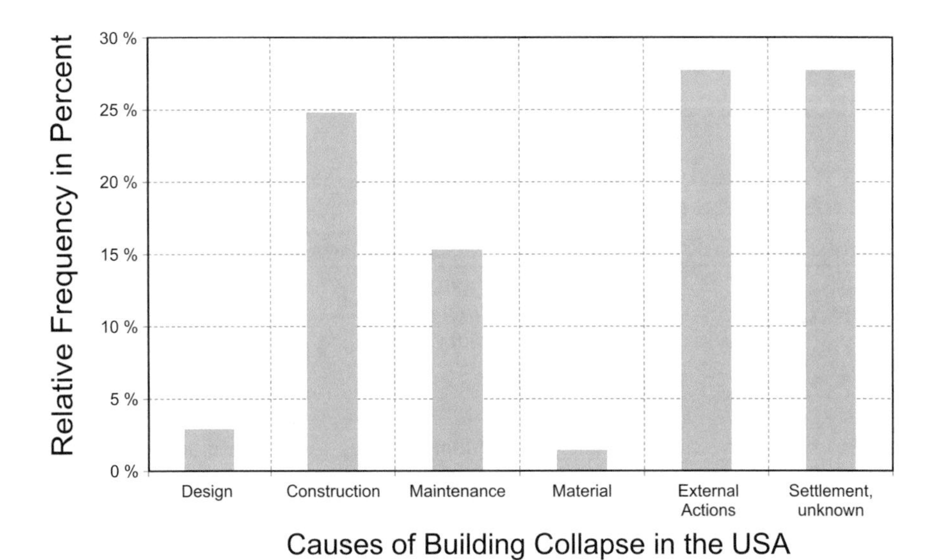

Fig. 7.6 Causes of building collapses according to Wardhana & Hadipriono (2003)

7.8 Mortality and Casualty Figures

This chapter does not consider deliberate collapses, such as the 2001 attack on the World Trade Centre in the USA, nor collapses caused by large-scale accidental actions such as floods, earthquakes, or war. For example, 400,000 houses were damaged in the 1970 Bangladesh flood (Frank & Husain 1971), 46,000 to 100,000 buildings were destroyed during the Kobe Earthquake 1995, 300,000 buildings were destroyed during the Tsunami-disaster 2004 (Zimmermann 2013), between 45,000 (Norio et al. 2011) and 130,000 (Kazama & Noda 2012) buildings were completely destroyed and between 190,000 (Norio et al. 2011) and 240,000 (Kazama & Noda 2012) buildings were partially destroyed in the 2011 Tohoku earthquake, and approximately 250,000 buildings were damaged in the 2010 Haiti earthquake (Bilham 2010). In the Second World War, almost 4 million buildings were destroyed in Germany alone (Hardinghaus 2020). Hiroshima lost about 60,000 of the city's 90,000 buildings to the atomic bomb in August 1945 (Groves 1946).

The calculation of the annual mean number of victims from earthquakes results in 17,000 based on the 20 most severe earthquakes (Proske 2020). Daniell et al. (2011) give a total number of victims from earthquakes for the twentieth century of 2.42 million yielding to 24,000 victims per year. According to Guha-Sapir & Vos (2011), the global mean annual number of victims from earthquakes between 1990 and 2011 was about 27,000.

Nichols & Beavers (2008) and Guha-Sapir et al. (2016) cite significantly smaller (about 9,000 per year) and significantly larger (35,000 per year) annual mean numbers of victims. Holzer & Savage (2013) estimate 2.57 million deaths from earthquakes for the twenty-first century. This corresponds to an annual mean number of victims of 25,700.

Nichols & Beavers (2008) estimate a mean number of victims per earthquake of about 400, pointing to a deficit in earthquake reports before 1900.

Approximately 75% of earthquake fatalities are caused by the collapse of structures (Coburn et al. 1992). Bilhalm (2009) estimates the cumulative number of victims of all earthquakes at more than 10 million people.

In addition to direct victims from building collapses, this affects certain professions. For example, 20% of all occupational deaths in the US fire service are due to building collapses (Stroup & Bryner 2007).

Figure 7.7 shows the evaluation and calculation of mortalities with data from literature and daily press (Lyu et al. 2018, Feifei 2014, National Chrime Records Bureau 2020, Omenihu et al. 2016, Danso & Boateng 2013, Asante & Sasu 2018, SKV Feuerwehr Wien 2020, Eldukair & Ayyub 1991, Reid 2000, Tagesanzeiger 2011, Windapo & Rotimi 2012, Blockley 1980, Rackwitz 1998, Tageblatt Frauenfeld 2017, Boateng 2012, RP Online I 2020, Liebers 1996, Wardhana & Hadipriono 2003 and for comparison purposes Vogel et al. 2009).

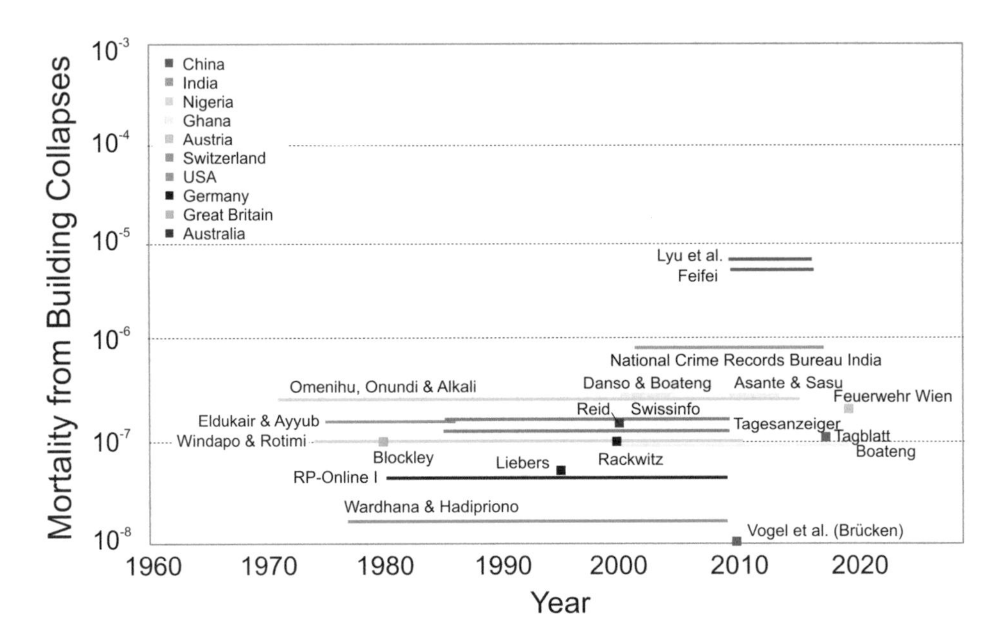

Fig. 7.7 Mortalities due to building collapses according to different authors (Proske & Schmid 2021)

Figure 7.8 summarises the curves by averaging countries and indicating developed and developing countries.

In the literature, mortalities due to building collapses are mentioned in various places. Blockley (1980), Rackwitz (1998) and Melchers (1999) mention a value of 10^{-7} per year. This value is also used by Haldi & Vulliet (1998), Imhof (2004) and Proske (2009). Reid (2000) and Canisius (2015) give a slightly higher value of 1.4×10^{-7} per year. Proske & Schmid (2021) calculate a mean annual mortality for developed countries of 9.8×10^{-8}, which is roughly equivalent to 10^{-7}, and of 1.5×10^{-6} for developing countries. The value for developing countries exceeds the known values from the literature by more than a factor of 10.

Mortality calculations of building collapse for certain accidental actions, e.g. explosions, are known (e.g. Hingorani et al. 2020), but were not used here.

Regions with similar populations, geological characteristics, and earthquakes but different building standards were compared. This shows significantly smaller numbers of victims for the developed countries (Table 7.4). In contrast, the ratio of injured to fatalities has increased (Table 7.5). This means that an improvement in the design and construction of buildings results in a shift away from fatalities towards injuries.

A calculation of the Fatal Accident Rate, i.e. the mortality normalised to the exposure time, does not actually have to be carried out for the risk from building collapses because the exposure time almost corresponds to the calendar time (Melchers 1999; Maag 2004).

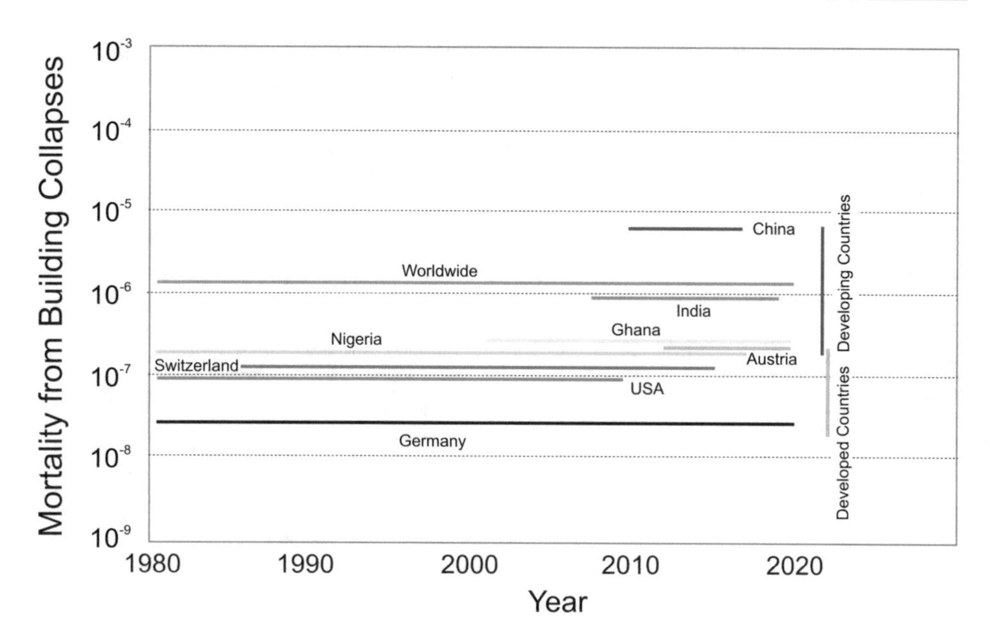

Fig. 7.8 Annual mortalities due to building collapses summarised for different countries (Proske & Schmid 2021)

Table 7.4 Victim and damage data of the comparable earthquakes of the Spitak earthquake in Armenia and the Loma Prieta earthquake in Northern California (Bachmann 2002, 1997; Newson 2001)

	Spitak-Earthquake	Loma-Prieta-Earthquake
Date	7. December 1988	17. October 1989
Magnitude	6.9	7.1
Region	Armenia	Northern California
Fatalities	>25,000	67
Injured	31,000	2,435
Homeless	514,000	7,362
Porperty Damage	unknown	approx. 10 Billion Swiss Francs

Recent recalculations of "years of life lost" due to building collapses show values increased by about 20% compared to previous studies: from 0.19 to 0.20 days of life to 0.255 days of life or from 4.56 to 4.8 h to 6.12 h. Whether this is a systematic change, especially due to the massive construction activity in developing countries, must be shown by further studies.

Table 7.5 Ratio of injured to fatalities in earthquakes of at least magnitude 6 and with at least 40 fatalities. Except for the first row (world), only shallow earthquakes on land were considered (Wyss & Trendafiloski 2009)

	500–1899	1900–1949	1950–1969	1970–1985	1985–2008
World	1.2	2.8	5.4	4.3	6.9
Developing Countries without China			3.0	3.2	4.8
Developed Countries without Japan			8.8		11.2
China			2.5		12.8
Japan			6.6		47.5
Latin America				2.6	8.0
Turkey, Iran			2.6		3.6
Greece			18.6		11.2
Italy			3.9		7.0

Table 7.6 Various risk values for building structures and buildings (Blockley 1980; Menzies 1996; Eurocode 0 2017; Proske 2018, 2020, 2022, 2009)

Risk parameters	Buildings and Structures	Target Value
Mortality per year	1×10^{-7}	10^{-6}
Fatal Accident Rate	0.002	0.2 to 2.0
Loss of years of life	4.8 h	For comparison reasons Cars: 200 days

The concept of years of life lost considers injuries as well as fatalities. A medical discussion of fatal and non-fatal injuries due to building collapses is given e.g. in Ikuta et al. (2004) and Yamazaki et al. (1996).

Table 7.6 summarises various risk parameters for life and limb due to building collapses.

7.9 Summary

Based on the study presented here, various conclusions can be drawn:

- The observed collapse frequency without large-scale accidental actions of 2.4×10^{-7} per year for developed countries is lower than the target values of failure probability per verification equation of 10^{-6} per year.

- The observed collapse frequency without large-scale accidental actions is 4.7×10^{-6} per year for developing countries, which is greater than the target failure probability per verification equation of 10^{-6} per year. Correction factors to account for correlation of the multiple structural analysis equations and for human error in developed countries suggest an increase in the target failure probability by a factor of 1.2 to 1.3 (Proske 2018). Thus, the value of 4.7×10^{-6} is still above the target value. Based on the comparison of Figs. 7.5 and 7.6, the question arises whether the factor must be higher in developing countries and whether the assumptions of modern construction standards regarding the exclusion of human error are thus realistic.
- The calculated mean failure probabilities with or without consideration of large-scale accidental actions are clearly above the observed collapse frequencies. The difference is significantly greater than for all other types of structures investigated so far.
- The mortality values for developed countries agree very well with other references. However, the values for developing countries are more than a power of ten higher.
- Building collapses without large-scale accidental actions reach between a half and a third of the global annual number of victims of building collapses due to large-scale accidental actions. Bilham (2009) estimates that more than ten million people have died from earthquake-induced building collapses throughout human history. Based on the conclusion formulated here, up to five million people would have died from building collapses throughout human history.
- Despite these considerable numbers of victims, risk values considering the exposure time and the "years of life lost" show that buildings are extraordinarily safe compared to other technical products. There may be various reasons for this, such as the large period required to rebuild the buildings, the lack of alternatives, the great importance for the human civilisations, the good accessibility for monitoring and the secondary function as a storage of wealth. The argument of great experience may only be valid to a limited extent because shipbuilding also has great experience and still shows significantly higher risk values.

References

Aagaard N-J, Pedersen ES (2013) Failure and documentation of building structures, CIB World Building Conress 2013, 5–9 May 2013, Birsbane, Australien

Agenzia nazionale (2016) Agenzia nazionale per le nuove tecnologie (2016) Portale4E, http://www.portale4e.it/pa_guide_dettaglio.aspx?ID=1

Allen DE, Schriever WR (1972) Progressive Collapse, Abnormal Loads and Building Codes, Structural Failures: Modes, Causes, Responsibilities — a compilation of papers printed at the ASCE National Meeting on Structural Engineering, Cleveland, Ohio, April 1972, published by ASCE, New York

Asante LA, Sasu A (2018) The challenge of reducing the incidence of building collapse in ghana: Analyzing the perspectives of building inspectors in kumasi. SAGE Open 2018:1–12

Ayedun CA, Durodola OD, Akinjare OA (2011) An empiricial ascertainment of the causes of building failure and collapse in Nigeria, mediterranean. J. Soc. Sci 3(1):313–322

Ayodeji O (2011) An examination of the causes and effects of building collapse in Nigeria. J. Des. Built Environ 9:37–47

Azhahar N, Karim NA, Hassan SH, Eman J (2011) A study of contribution factors to building failures and defects in construction industry. Procedia Eng 20(2011):249–255

Bachmann H (1997) Erdbebensicherung der Bauwerke. In: Mehlhorn G (ed.) Der Ingenieurbau: Grundwissen, Teil 8: Tragwerkszuverlässigkeit, Einwirkungen. Verlag Wilhelm Ernst & Sohn, Berlin

Bachmann H (2002) Erdbebensicherung von Bauwerken, Birkhäuser Verlag, Basel, 2. Auflage

Baldegger J, Nathani C, Anderloni F, Bachmann,F, Kolb J, Mulle R, Brandes J, Hellmüller P (2020) Die volkswirtschaftliche Bedeutung der Immobilienwirtschaft der Schweiz, Hauseigentümerverband Schweiz und Bundesamt für Wohnungswesen (Hrsg.), Rüschlikon, März 2020

Baunetz.de (1998) Tragisches Pech am Bau, 26. Juni 1998, https://www.baunetz.de/meldungen/MeldungenSporthalle_Halstenbek_zum_zweiten_Mal_eingestuerzt_mit_Kommentar_der_Redaktion__3805.html

Behnisch M, Meinel G, Burckhardt M, Hecht R (2012) Auswertungen zum Gebäudebestand in Deutschland auf Grundlage digitaler Geobasisdaten. In: Meinel G, Schumacher, U & Behnisch M (Hrsg.): Flächennutzungsmonitoring IV. Genauere Daten – informierte Akteure – praktisches Handeln. Berlin: Rhombos, IÖR Schriften 60, pp. 151–158, http://www.ioer-monitor.de/fileadmin/user_upload/monitor/DFNS/2012_4_DFNS/Buchbeitraege/IOER_Schrift_60_DFNS_2012_S151-158_A.pdf

Bilham R (2009) The seismic future of cities, Bull Earthquake Eng, November 2009, 50 pages

Bilham R (2010) Lessons from the Haiti Earthquake, *Nature* 463(18):878–879

Blockley DI (1980) The nature of structural design and safety, 1980. Wiley & Sons, Chichester

Boateng FG (2012) A critique of overpopulation as a cause of pathologies in African cities: Evidence from building collapse in Ghana, World Development, 137, January 2021, doi.org/https://doi.org/10.1016/j.worlddev.2020.105161

BPIE (2011) European Buildings under the microscope, A country-by-country review of the energy performance of buildings, Building Performance Institute Europe, Brüssel

Bundesamt für Statistik (2019) Gebäude, https://www.bfs.admin.ch/bfs/de/home/statistiken/bau-wohnungswesen/gebaeude.html

Canisius G (2015) Robustness of Structures, October 2015, WSP/Parsons Brinckerhoff, Presentation, with Material from H Gulvanessian

CBECS (2015) A look at the U.S. commercial building stock: Results from EIA's 2012 commercial buildings energy consumption survey (CBECS), 4. März 2015

Coburn AW, Spence RJS, Pomonis A (1992) Factors determining human casualty levels in earthquakes: Mortality prediction in building collapse. Earthquake Engineering, Tenth World conference, Balkema, Rotterdam, 5989–5994

Daniell JE, Khazai B, Wenzel F, Vervaek A (2011) The CATDAT damaging earthquakes database. Nat Hazards Earth Syst Sci 11:2235–2251

Danso H, Boateng I (2013) Is the Quality of Cement a Contributing Factor for Building Collapse in Ghana? In: Laryea, S. and Agyepong, S (Eds.), Proceedings of the 5th West Africa Built Environment Research (WABER) Conference, 12–14 August 2013, Accra, Ghana, 765–772

Dena (2019) dena-Gebäudereport Kompakt 2019, Statistiken und Analysen zur Energieeffizienz im Gebäudebestand, Deutsche Energie-Agentur GmbH (dena), Berlin

Destatis (2019) Anzahl der Wohngebäude in Deutschland in den Jahren 2000 bis 2017, https://de.statista.com/statistik/daten/studie/70094/umfrage/wohngebaeude-bestand-in-deutschland-seit-1994/

DSW (2019) Soziale und demografische Daten weltweit, DSW Datenreport 2019, Deutsche Stiftung Weltbevölkerung (DSW), Hannover

Duden (2016) Wirtschaft von A bis Z: Grundlagenwissen für Schule und Studium, Beruf und Alltag. 6. Aufl. Mannheim: Bibliographisches Institut 2016. Lizenzausgabe Bonn: Bundeszentrale für politische Bildung

Ede AN, Olofinnade OM (2018) Awoyera PO (2018) Structural form works and safety challenges: Role of Bamboo Scaffold on collapse of reinforced concrete buildings in Nigeria. Int. J. Civ. Eng. Techn 9(9):1675–1681

Eldukair ZA, Ayyub BM (1991) Analysis of recent U.S structural and construction failures. J. Perform. Constr. Fac 4:57–73

Emuze F, van Eeden L, Geminiani F (2015) Causes and Effects of Building Collapse: A case Study in South Africa, Proceedings of CIB W099, Benefitting Workers and Society through Inherently Safe(r) Construction, Belfast, Northern Ireland, 10–11 September 2015, 408–417

Ernst S (2006) Lagos, Hyperwachstum – ungebremst und informell, Bundeszentrale für politische Bildung, 19. Oktober 2006, https://www.bpb.de/internationales/weltweit/megastaedte/64606/lagos?p=all

Eurocode 0 (2017) EN 1990 Basis of Structural Design, 2nd Edition, Draft 30 April 2017

Fabbri M (2020) Clean Energy Package: why buildings matter, Presentation Build-ings_101_PBIE.pptx

Fagbenle O, Oluwunmi AO (2010) Building failure and collapse in Nigeria: The influence of the informal sector. J. Sustain. Dev 3(4):268–276

Feifei F (2014) Overview of building collapses in China, Chinadaily, 4 April 2014 https://www.chinadaily.com.cn/china/2014-04/04/content_17408943.htm

Feuerwehr Trieben Stadt (2013) Gebäudeeinsturz, 16. Juli 2013, https://www.ff.at/trieben/bericht/252

Frank NL, Husain SA (1971) The deadliest cyclone in history? Bull Am Meteor Soc 52(6):438–445

Freistaat Sachsen (2018) Sächsische Bauordnung in der Fassung der Bekanntmachung vom 11. Mai 2016 (SächsGVBl. S. 186) mit letzter Änderung vom 11. Dezember 2018 (Sächs-GVBl. p. 706)

Geis J (2012) Snow-Induced Building Failures, American Society of Civil Engineers

GGS (2013) 2010 Population & Housing Census, National Analytical Report, Ghana Statistical Service (GSS), May 2013

Government of Canada (2020) The open database of buildings, https://open.canada.ca/data/en/dataset/40e37a0f-1393-4e91-bd00-334dceb26e34

Griess A (2015) Bauboom in China, 31. März 2015 https://de.statista.com/infografik/3357/flaeche-von-gebaeuden-im-bauzustand-in-china/

Groves LR (1946) The atomic bombings of Hiroshima and Nagasaki, Manhattan Engineer District of the United States Army, June 29, 1946, https://www.atomicarchive.com/resources/documents/med/index.html

GTAI (2020) Germany Trade & Invest GTAI: Tschechien fördert Energieeffizienz bei Neu- und Altbauten, 9. April 2020, https://www.gtai.de/gtaide/trade/branchen/branchenbericht/tschechische-republik/tschechien-foerdertenergieeffizienz-bei-neu-und-altbauten-23012

Guha-Sapir D, Hoyois P, Wallemacq P, Below R (2016) Annual disaster statistical review 2016: The numbers and trends. Centre for Research on the Epidemiology of Disasters, Brussels

Guha-Sapir D, Vos F (2011) Earthquakes, an Epidemiological Perspective on Pattern and Trends, In: Human Casualties in Earthquakes, Progress in Modelling and Mitigation, R. Spence, E So and Scawthorn, Springer: Dordrecht – Heidelberg – London – New York, pp. 13–24

Hadipriono FC (1985) Analysis of events in recent structural failures. J. Struct Eng 111(7):1468–1481

Haldi PA, Vulliet L (1998) Fiabilité et sécurité des systèmes civils. Lecture notes, Swiss Federal Institute of Technology (EPFL), Lausanne

Hamma-adama M, Kouider T (2017) Causes of building failure and collapse in Nigeria: Professionals view. Am. J. Eng. Res (AJER) 6(12):289–300

Hardinghaus C (2020) Die Verratene Generation - Gespräche mit den letzten Zeitzeuginnen des Zweiten Weltkrieges, Europaverlag Zürich

HEV (2016) Schweiz: Wohneigentum in Zahlen, 2015, https://www.hev-schweiz.ch/fileadmin/sektionen/hev-schweiz/PDFs_Dateien/Jahresberichte/HEV_Leporello_2015.pdf

Hingorani R, Tanner P, Prieto M, Lara C (2020) Consequence classes and associated models for predicting loss of life in collapse of building structures. Struct Saf 85:1–13

Holzer TL, Savage JC (2013) Global earthquake fatalities and population, earthquake. Spectra 29(1):155–175

Hong L, Zhou N, Feng W, Khanna N, Fridley D, Zhao Y, Sandholt K (2016) Building stock dynamics and its impacts on materials and energy demand in China, *Energy Policy* 94(7), 47–55, https://www.hev-schweiz.ch/fileadmin/sektionen/hev-schweiz/PDFs_Dateien/Jahresberichte/HEV_Leporello_2015.pdf

Ikuta E, Miyano M, Nagashima F, Nishimura A, Tanaka H, Nakamori Y, Kajiwara K, Kumagai Y (2004) Measurement of the Human Body Damage caused by Collapsed Building, 3th World Conference on Earthquake Engineering Vancouver, Canada, August 1–6 2004, Paper No. 628, 13 pages

Imhof D (2004) Risk Assessment of Existing Bridge Structures, University of Cambridge, Dissertation, Kings College, December 2004

Kaul V (2015) Why are more than 10 million homes vacant in India? BBC News, 21. May 2015, https://www.bbc.com/news/world-asia-india-32644293

Kazama M, Noda T (2012) Damage statistics (Summary of the 2011 off the Pacific Coast of Tohoku Earthquake damage). Soils Found 52(5):780–792

Krausmann F, Wiedenhofer D, Lauk Chr, Haas W, Tanikawa H, Fishman T, Miatto A, Schandl H, Haberl H (2017) Global socioeconomic material stocks rise 23-fold over the 20th century and require half of annual resource use, PNAS February 21 2017, Vol. 114, Issue 8, pp. 1880–1885; first published February 6, https://doi.org/10.1073/pnas.1613773114

Liebers P (1996) Jenas Roter Turm Fall für Gerichtshof in Luxemburg, Neues Deutschland, 24 April 1996, https://www.neues-deutschland.de/artikel/606607.jenas-roter-turm-fall-fuergerichtshof-in-luxemburg.html

Luzernerzeitung (2019) Fünf Verletzte bei Deckeneinsturz in Bar in Bellinzona, 12. Februar 2019, https://www.luzernerzeitung.ch/newsticker/schweiz/funf-verletzte-bei-deckeneinsturz-in-barin-bellinzona-ld.1093533,

Lyu H-M, Cheng W-C, Shen JS, Arulrajah A (2018): Investigation of Collapsed Building Incidents on Soft Marine Deposit: Both Social and Technical Perspectives, Land, Vol. 7, Issue 20, 12 pages

Maag T (2004) Risikobasierte Beurteilung der Personensicherheit von Wohnbauten im Brandfall unter Verwendung von Bayes`schen Netzen, Doctorial Thesis, IBK Bericht, vol. 282, Zürich: vdf Hochschulverlag AG an der ETH Zürich

Malo J (2018) All of Australia's 15.2 million buildings have been mapped, 31. Oktober 2018, https://www.domain.com.au/news/ai-machine-learning-helped-a-canberracompany-map-every-building-in-australia-779281

Mann C (2011) Die Geburt der Zivilisation, National Geographic Deutschland, Juni 2011, Pp. 38
Mann G (1991) Propyläen Weltgeschichte - Eine Universalgeschichte. Band 8. Propyläen Verlag Berlin - Frankfurt am Main
meinbezirk.at (2012) Gebäudeeinsturz in Prutz, 19. Februar 2012, https://www.meinbezirk.at/landeck/c-lokales/gebaeudeeinsturz-in-prutz_a137989
Melchers RE (1999) Structural reliability analysis and prediction. John Wiley & Son Ltd.
Menzies JB (1996) Bridge Failures, Hazards and Societal Risk, International Symposium on the Safety of Bridges, July 1996, London
Michael AO, Razak AR (2013) The Study of Claims Arising from Building Collapses: Case Studies from Malaysia, Nigeria, Singapore and Thailand, Civil and Environmental Research, 3(11), 113–128
National Building Organisation (2016) Slums in India, A Statistical Compendium 2015, Ministry of Housing and Ruban Poverty Alleviation, Government of India, New Delhi, 2016
National Crime Records Bureau (2020) Accidental Deaths & Suicides in India 2019, Ministry of Home Affairs, Neu Dehli
Naviant Research (2018) Global Building Stock Database - Commercial and Residential Building Floor Space by Country and Building Type: 2017-2026, Chicago
Newson L ((2001) The Atlas of the World's Worst Natual Disasters. Dorling Kindersley, London
Nichols JM, Beavers JE (2008) World earthquake fatalities from the past: Implifications for the present and future. Nat. Hazards Rev. 9(4):179–189
Norio O, Ye T, Kajitani Y, Shi P, Tatano H (2011) The 2011 eastern Japan great earth-quake disaster: Overview and comments. Int. J Disaster Risk Sci 2(1):34–42
Omenihu FC, Onundi LO, Alkali MA (2016) An Analysis of Building Collapse in Nigeria (1971–2016): Challenges for Stakeholders, University of Maiduguri, Annals of Borno, Volume XXVI, June 2016, pp. 113,139
Openstreetmap (2019) Canada/Building Canada 2020, 23. Juli 2019 https://wiki.openstreetmap.org/wiki/Canada/Building_Canada_2020
ORF Tirol (2015) Balkon in Völs eingestürzt, 28. Juli 2015, https://tirol.orf.at/v2/news/stories/2723580/
Ortega I (2000) Systematic Preventation of Construction Failures, ITEM-HSG, Report No. 9, Report Series Quality Management and Technology, University of St. Gallen, Januar 2000, pp. 1–13
Proske D (2009) Catalogue of Risks, Springer: Heidelberg – New York
Proske D (2018) Bridge collapse frequencies versus failure probabilities. Springer, Cham
Proske D (2020) Die globale Gesundheitsbelastung durch Bauwerksversagen, Bautechnik, 97(4), 233–242
Proske D (2021) Estimation of the Global Health Burden of Structural Collapse. Matos, J.C., Lourenço, P.B., Oliveira, D.V., Branco, J., Proske, D., Silva, R.A., Sousa, H.S. (Eds.) 18th International Probabilistic Workshop, Lecture Notes in Civil Engineering, Springer International Publishing, pp. 327–340
Proske D (2022) Katalog der Risiken, 2. vollständig überarbeite Auflage, Spinger
Proske D, Schmid M (2021) Häufigkeit von und Mortalität bei Hochbaueinstürzen, Bautechnik, 98(6):423–432.
Proske D, Schmid M (2022) Comparison of the Collapse Frequency and Failure Probability of Buildings, Proceedings of the International Probabilistic Workshop, September 2022, Stellenbosch, South Africa, in press
Rackwitz R (1998) Zuverlässigkeit und Lasten im konstruktiven Ingenieurbau, Teil I: Zuverlässigkeitstheoretische Grundlagen, Technische Universität München, 1993–1998
Reid SG (2000) Acceptable risk criteria. Prog Struct Mat Eng 2(2):254–262

RP Online I (2020) Chronik: Spektakuläre Einstürze in Deutschland, https://rponline.de/panorama/deutschland/chronik-spektakulaere-einstuerze-in-deutschland_iid12281877

RP Online II (2005) Drei Tote bei Balkon-Einsturz in Brühl, 29 Juli 2005, https://rponline.de/panorama/deutschland/drei-tote-bei-balkon-einsturz-in-bruehl_aid-9037363

Scheer J (2001) Versagen von Bauwerken, Band 2, Hochbauten und Sonderbauwerke, Ernst und Sohn, September 2001, Berlin

Schiller G, Ortlepp R, Krauß N, Steger S, Schütz H, Fernández JA (2015) Reichenbach J, Wagner J & Baumann J: Kartierung des anthropogenen Lagers in Deutschland zur Optimierung der Sekundärrohstoffwirtschaft, Umweltbundesamt, Dessau-Roßlau, Juli 2015

Schweizerische Eidgenossenschaft (2017) Verordnung über das eidgenössische Gebäude- und Wohnregister (VGWR), Bern, 2017, vom 9. Juni 2017

Sirviö A, Illikainen K (2015) Sustainable Buildings for the High North. Strategic plan for the implementation of energy efficient renovation and construction in Northern parts of Russia, ePooki 28/2015, 9. Dezember 2015

SKV Feuerwehr Wien (2020) Einsätze, http://www.feuerwehrwien.at/html/einsaetze.html.

Statista (2020) Number of Housing Units in the United States, https://www.statista.com/statistics/240267/number-of-housing-units-in-the-unitedstates/, 13 Mai 2020. https://www.statista.com/statistics/240267/number-of-housing-units-in-the-united-states/

Statistik Austria (2015) Bundesanstalt Statistik: Bestand an Gebäuden und Wohnungen, 16 März 2015, https://www.statistik.at/web_de/statistiken/menschen_und_gesellschaft/wohnen/wohnungs_und_gebaeudebestand/index.html

Statistisches Bundesamt (2014) Systematik der Bauwerke, Erstausgabe 1978, Version vom 1.1.2014, Wiesbaden

Stroup DW, Bryner NP (2007) Structural Collapse Research at NIST, 13 pages

Stuttgarter Nachrichten (2019) Unterschätzte Gefahr auf dem Balkon, 1 Juli 2019, https://www.stuttgarter-nachrichten.de/inhalt.einsturz-in-stuttgart-unterschaetztegefahr-auf-dem-balkon.cc396eb2-4c58-4040-b94a-228ac36a7230.html. Zugegriffen 8 Okt. 2020

Sulaima MF, Lew HS, Lau CY, Lim CKY, Azily AT (2014) A Case Study of Engineering Ethics: Lesson Learned from Building Collapse Disaster toward Malaysian Engineers, European International Journal of Science and Technology, Vol. 3, No. 4, May 2014, pp. 21-30

Swissinfo (2004) Seit Uster der folgenschwerste Einsturz, 28. November 2004, https://www.swissinfo.ch/ger/seit-usterder-folgenschwerste-einsturz/4222880

Tagblatt Frauenfeld (2017) „Obstgarten"-Einsturz: „Tragisch, was passiert ist", 17. November 2017, https://www.tagblatt.ch/ostschweiz/frauenfeld-munchwilen/frauenfeld-obstgarten-einsturztragisch-was-passiert-ist-ld.767369

Tagesanzeiger Panorama (2011) Die schlimmsten Deckeneinstürze in der Schweiz, 13. April 2011, https://www.tagesanzeiger.ch/panorama/vermischtes/die-schlimmsten-deckeneinstuerze-derschweiz/story/21910073.

Tanikawa H, Fishman T, Okuoka K, Sugimoto K (2015) The weight of society over time and space - A comprehensive account of the construction material stock of Japan, 1945–2010. J Ind Ecol 9(5):778–791

United Nations (2018) Tracking Progress Towards Inclusive, Safe, Resilient and Sustainable Cities and Human Settlements. DG 11 Synthesis Report 2018, Nairobi, Kenya

Vogel T, Zwicky D, Joray D, Diggelmann M, Hoj NP (2009) Tragsicherheit der bestehenden Kunstbauten, Sicherheit des Verkehrssystems Strasse und dessen Kunstbauten. Federal Roads Office (ASTRA), 12/2009, Bern

Wardhana K, Hadipriono FC (2003) Study of Recent Building Failures in the United States, Journal of Performance of Construction Facilities, ASCE, August 2003, pp. 151–158

Windapo AO, Rotimi JO (2012) Contemporary Issues in Building Collapse and Its Implications for Substainable Development, Buildings, Vol. 2, Issue 4, 25. Juli 2012, pp. 283–299

Wyss M, Trendafiloski G (2009) Trends in the Casualty Ration of Injured to Fatalities in Earthquakes. Second International Workshop on Disaster Casualties, 15–16 June 2009, University of Cambridge, UK, pp. 1–6

Yamazaki F, Nishimura A, Ueno Y (1996) Estimation of Human Casualties Due to Urban Earthquakes, Eleventh World Conference on Earthquake Engineering, Elsevier Science Ltd, 8 pages

Zerna W (1983) Grundlage der gegenwärtigen Sicherheitspraxis in der Bautechnik, In: Große technische Gefahrenpotentiale: Risikoanalysen und Sicherheitsfragen, Hrsg. S. Hartwig, Pp. 99–109

ZIA (2017) Zentraler Immobilien Ausschuss e.V.: Immobilienwirtschaft 2017, Berlin

Zimmermann I (2013) Entwicklung von Strategien für die Konstruktion von tsunamisicheren Bauweisen. Technische Universität Braunschweig, Braunschweig, August, Bachelorarbeit

Stadiums

8

8.1 Introduction

Stadiums are a special type of buildings and structures. The construction of stadiums has a long history. The Colosseum in Rome is particularly famous, but many other cities, such as Delphi and Olympia, also had and still have historic stadiums. Stadiums have therefore been built for at least 2,500 years.

8.2 Definition of Stadiums

In general, stadiums are structures with the purpose of allowing large groups of people to gather for arts or sporting events, while ensuring visibility of the stage or sports field and often providing weather protection for visitors. Stadiums usually have a central area for the cultural or sporting event, which is usually open to the top.

However, examples of further definitions are known and given here:

The German Duden (Dudenredaktion 2015) defines stadiums as "*large facilities with stands for spectators for sporting competitions and exercises, especially in the form of a large, often oval sports field.*"

According to Gorokhov et al. (2013), modern stadiums are characterised by their universality in the possibilities of staging sporting events and cultural programs. In addition, they have the task of providing maximum comfort to spectators.

Pfaff (2000) considers stadiums to mean "*a multi-purpose sports facilities equipped with a football pitch and a 400 m track (incl. jumping and throwing facilities)*". Stadiums originated from amphitheatres. However, Pfaff (2000) sees fundamental differences in the characteristics of arenas and stadiums. Arenas are more experience-oriented facilities without tracks and offer many seats as well as VIP boxes. In contrast, stadiums are

D. Proske, *The Collapse Frequency of Structures*,
https://doi.org/10.1007/978-3-030-97247-9_8

facilities equipped with tracks and standing room, where the focus is on the performance of sporting activities.

8.3 Stock of Stadiums

The Stadium Guide (2020) estimates that there are about 5,000 large stadiums worldwide (Table 8.1) but own extrapolating from major European cities yields higher figures. For example, the city of Dresden in Germany has a football stadium, an athletics stadium, an ice hockey stadium, a horse racing track, and numerous smaller sports fields. The city has about 500,000 inhabitants. If one estimates that there are about 250,000 inhabitants per stadium and apply this to about 1.5 billion people in industrialized countries, one receives about 6,000 stadiums. However, the large number of stadiums in cities does not apply to rural areas. If one assumes that about 2/3 of the population lives in cities, one receives 4,000 stadiums. Since there are also numerous large stadiums in Asia and South America, one could assume: 7.5 billion people × 0.5 for the proportion in cities × 1 / 500,000 = 7,500 large stadiums. The number of large stadiums worldwide is therefore probably between 5,000 and 7,500.

8.4 Calculation of Collapse Frequencies

8.4.1 Introduction

For the computation of the collapse frequency five publication were used (van Vliet 2016, Stadium Guide 2020, BBC.com 2020, CBS-News 2012, Wikipedia 2020).

Table 8.1 Number of stadiums according to Stadium Guide (2020)

Continent	Number of stadiums
Africa	402
Asia	1,281
Central America	86
Europe	1,244
Middle East	316
North America	927
Australia and Oceania	121
South America	479
World	4,855

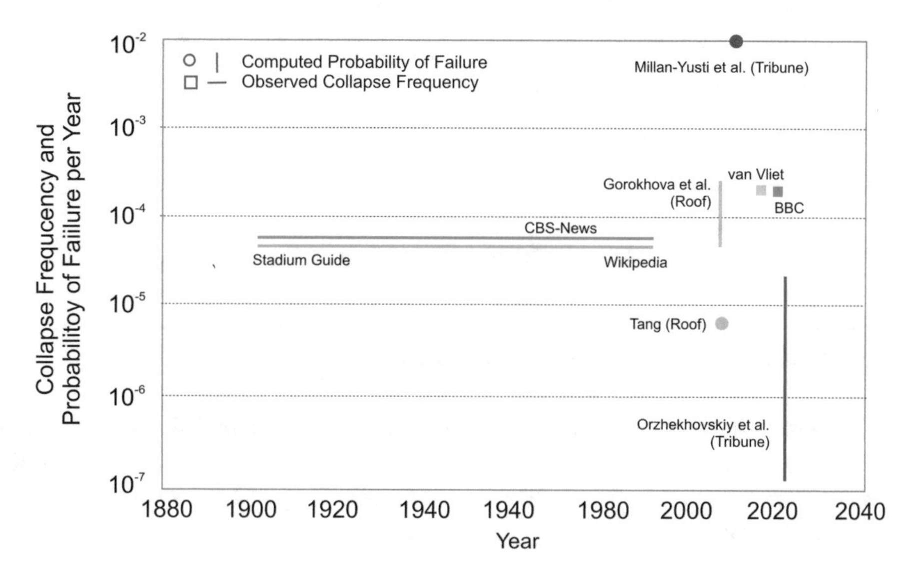

Fig. 8.1 Collapse frequencies and failure probabilities of stadiums

Table 8.2 Mean values and standard deviations

Probability of failure		Collapse frequency		Target value per year
Standard deviation	Mean value	Standard deviation	Mean value	
2.5	2.8×10^{-4}	0.44	8.41×10^{-5}	10^{-6}

Examples of stadium collapses or stadium roof collapses can also be found in expert reports, such as NN (2009) and Department of Building and Housing (2012) (Fig. 8.1).

8.4.2 Central Estimator

The observed collapse frequency is above the collapse frequency for buildings in general, but still below 10^{-4} per year. For estimation of the mean value of the failure probability, the extreme value of 10^{-2} per year has given a lower weight (Table 8.2).

8.4.3 Measure of Deviation

The measure of deviation, here in terms of standard deviation, is extremely low for the observed collapse frequencies. This may be related to the low number of publications, the low number of structures and the limited number of collapses. The different references refer often to the same collapse events.

8.4.4 Trend

Due to the low number of references no trend is investigated.

8.5 Calculation of Failure Probabilities

Four references have been used to investigate the computed probabilities of failure (Gorokhov et al. 2013, Tang 2013, Millan-Yust et al. 2016, Orzhekhovskiy et al. 2020). There are other publications know for the computation of sport halls such as Sykora & Holicky (2009) but they have not been considered.

8.6 Comparison

While the mean values of the observed collapse frequencies and the calculated failure probabilities fit very well, the standard deviations are significantly different. The ranges of the samples are between one order of magnitude versus five orders of magnitude.

8.7 Causes

Without presenting detailed investigations here, it is known that stadiums can experience considerable actions due to the frequently used lightweight construction and the dynamic effects from wind and visitors. These two effects also dominate the causes of failure. (Wimmer 2016; Barnes & Dickson 2000).

8.8 Summary

Overall, only a few samples are available currently, so that a conclusive evaluation is not robust. The observed collapse frequencies are very close to each other - about one power of ten - which can also be attributed to the use of identical collapse events, but the calculated failure probabilities show a disproportionately large scatter of five powers of ten. In addition, the collapse frequencies cover a period of 100 years, while the failure probabilities only cover a period of about 10 years. Further investigations are necessary here.

References

Barnes M, Dickson M (2000) Widespan roof structures. Thomas Telford Ltd., London

BBC.com (2020) Worker dies in Russian sport stadium roof collapse. https://www.bbc.com/news/world-europe-51331181

CBS-News (2012) Major soccer stadium disasters. https://www.cbsnews.com/news/major-soccer-stadium-disasters/

Department of Building and Housing (2012) Technical investigation into the collapse of the Stadium Southland roof, May 2012, Wellington, New Zealand

Dudenredaktion (2015) Duden: Deutsches Universalwörterbuch, 8th edn. Dudenverlag, Berlin

Gorokhov YV, Voldymyr MP, Pryadko IN (2013) Reliability provision of rod shells of steady roofs over stadium stands an stage of design work. Procedia Engineering 57:353–363. https://doi.org/10.1016/j.proeng.2013.04.047

Millan-Yusti DC, Marulanda J, Thomsson P (2016) Valuation of the structural reliability of a grandstand subjected to anthropic loads. Universidad Del Valle, Ingeniería y Competitividad 18(1):58–68. https://www.redalyc.org/jatsRepo/2913/291343439006/html/index.html

NN (2009) The collapse stadium roof. https://ktsadium.wordpress.com/2009/08/14/8/

Orzhekhovskiy A, Priadko I, Tanasoglo A, Fomenko S (2020) Design of stadium roofs with a given level of reliability. Eng Struct 209 (15 April 2020). https://doi.org/10.1016/j.engstruct.2020.110245

Pfaff SM (2000) Erlebniswelt Arena, Zur Vermarktung von modernen Sportveranstaltungsstätten. Diplomica Verlag GmbH, Hamburg

Stadium Guide (2020) Stadium Disasters. https://www.stadiumguide.com/timelines/stadium-disasters/

Sykora M, Holicky M (2009) Failures of roofs under snow load: causes and reliability analysis, Fifth Forensic Engineering Congress 2009, Pathology of the Built Environment, Washington DC, United States, pp 444–452. https://doi.org/10.1061/41082(362)45

Tang S (2013) Study on system reliability of large tensile structure based on failure of main cables. Adv Mater Res 753–755:1477–1482. https://doi.org/10.4028/www.scientific.net/AMR.753-755.1477

van Vliet E (2016) Stadium roof collapse failure analysis, https://www.engineeringclicks.com/failure-analysis-stadium-roof-collapse/

Wikipedia (2020) Fußballstadien, https://de.wikipedia.org/wiki/Liste_von_Katastrophen_in_Fu%C3%9Fballstadien

Wimmer W (2016) Stadium buildings - construction and design manual, May 2016. DOM publishers, Berlin

9 Wind Turbines

9.1 Introduction

Wind energy has been used purposefully by humans for almost 4,000 years (Harding and Harris 1980). The focus was often on pumping water, but also on processing food or mechanical engineering. In the middle of the nineteenth century, there were probably 200,000 windmills in Europe (IG Windkraft). However, these were replaced by motors. Wind turbines have been built since the end of the nineteenth century.

9.2 Definition of Wind Turbines

Wind turbines are tower-shaped constructions for generating electrical energy from the kinetic energy of the wind (IEC 61400–1 2019). Wind turbines are distinguished both by location (on land, at sea) and by size (small size, etc.) (IEC 61400–2 2013, IEC 61400–3 2009).

Wind turbines have a large share of mechanical systems. It has not been conclusively clarified whether the tower of a wind turbine is considered a component of a machine (mechanical engineering) or a structure (civil engineering).

9.3 Stock of Wind Turbines

The worldwide number of wind turbines is increasing rapidly due to the conversion to sustainable power generation. Table 9.1 lists the world stock numbers based on different references.

D. Proske, *The Collapse Frequency of Structures*,
https://doi.org/10.1007/978-3-030-97247-9_9

Table 9.1 Development of the stock of wind turbines

Year	Reference	Number of wind turbines
2011	Česká společnost pro větrnou energii (2011)	199,064
2013	C&M Machinery (2013)	241,100
2014	Comulada (2016)	268,000
2015	GWEC (2015)	314,000
2016	Frangoul (2017)	341,000

For the last years only the growth of electrical power is known. Based on the increase in electrical output and the increase in maximum output per wind turbine, the current estimate for the worldwide number is around 450,000 wind turbines.

9.4 Calculation of Collapse Frequencies

9.4.1 Introduction

For the computation of the collapse frequency three publication were used (Caithness Windfarm Information Forum 2021; Jui-Sheng & Wan-Ting 2011; Yang et al. 2019). The number of considered collapses per reference is between 48 and 206. The references consider a maximum time span from 1996 to 2020 (Fig. 9.1).

9.4.2 Central Estimator

The observed collapse frequency is 9.40×10^{-5} per year, which is close to 10^{-4} per year.

9.4.3 Measure of Deviation

The measure of deviation, here in terms of span, is extremely low for the observed collapse frequencies: less the one order of magnitude. This may be related to the low number of publications (Table 9.2).

9.4.4 Trend

Due to the low number of references no trend is investigated.

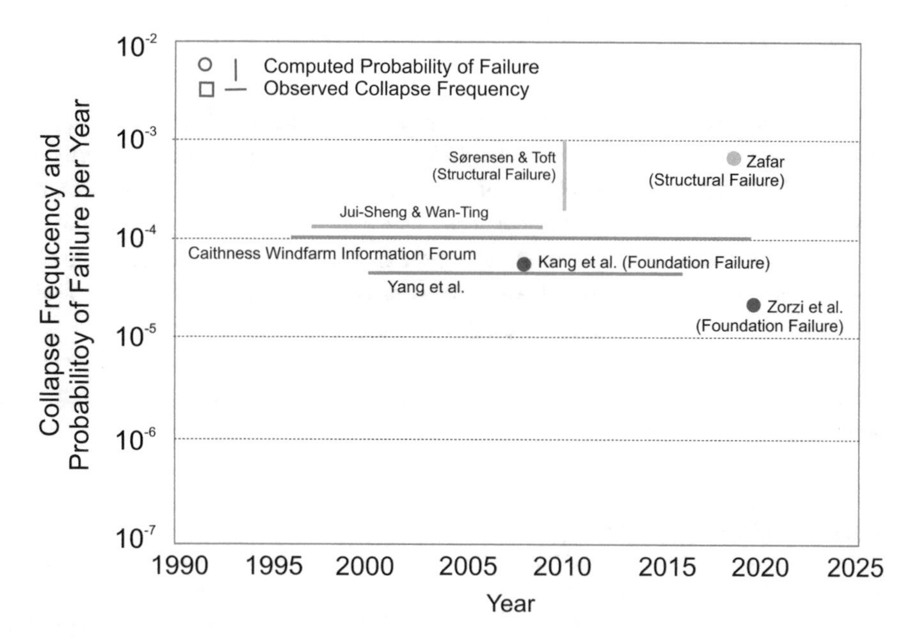

Fig. 9.1 Collapse frequencies and failure probabilities of stadiums

Table 9.2 Mean values and standard deviations

Probability of failure		Collapse frequency		Target value per year
Standard deviation	Mean value	Standard deviation	Mean value	
0.95	1.23×10^{-4}	0.5	9.40×10^{-5}	10^{-6}

9.5 Calculation of Failure Probabilities

Four references have been used to investigate the computed probabilities of failure (Kang et al. 2008, Sorensen and Toft 2010, Zafar 2019, Zorzi et al. 2020).

9.6 Comparison

Both, mean values and standard deviations of the observed collapse frequencies and the calculated failure probabilities fit very well.

9.7 Summary

Although only a small number of studies is available for both the observed collapse frequencies and the calculated failure probabilities, and although an extraordinarily high dynamic in the number and development of wind turbines can be observed, both the observed and the calculated values agree well. This conclusion applies to both the mean values and the deviation. It could be an indication that for newer constructions the observations and calculations agree better than for older constructions. Many types of structures, such as bridges, show a higher average age compared to wind turbines.

References

C&M Machinery (2013) New projects of Chines wind mills for electricity. https://www.domkiletniskowereda.pl/brazil/2020_Jul_22_2845/

Caithness Windfarm Information Forum 2021 (2021) Summary of wind turbine accident data to 31 March 2021. http://www.caithnesswindfarms.co.uk/AccidentStatistics.htm

Česká společnost pro větrnou energii (2011) Wind in numbers. https://csve.cz/en/clanky/wind-in-numbers/510

Comulada J (2016) 3 countries are capturing wind to power all our futures. https://www.upworthy.com/3-countries-are-capturing-wind-to-power-all-our-futures

Frangoul A (2017) There are over 341,000 wind turbines on the planet: Here's how much of a difference they're actually making. https://www.cnbc.com/2017/09/08/thereare-over-341000-wind-turbines-on-the-planet-why-they-matter.html#:~:text=Sustainable%20Energy-,There%20are%20over%20341%2C000%20wind%20turbines%20on%20the%20planet%3A%20Here's,difference%20they're%20actually%20making&

GWEC (2015) Wind in numbers. http://www.energybc.ca/cache/wind3/windnumbers.html

Harding E, Harris R (1980) The generation of electricity by wind power. Spon, London

IEC 61400–1 (2019) Windenergieanlagen Teil 1: Auslegungsanforderungen. International Electronical Commission, Fehraltorf

IEC 61400–2 (2013) Windenergieanlagen Teil 2: Anforderungen für kleine Windenergieanlagen. International Electronical Comission, Fehraltorf

IEC 61400–3 (2009) Windenergieanlagen Teil 3: Auslegungsanforderung für Windenergieanlagen auf offener See. International Electronical Comission, Fehraltorf

IG Windkraft (2021) Geschichte der Windkraft. https://www.igwindkraft.at/?xmlval_ID_KEY[0]=1045#:~:text=Die%20moderne%20Windkraftnutzung%20f%C3%BCr%20die%20Stromerzeugung%20beginnt%20knapp%20vor%201900%20in%20D%C3%A4nemark

Jui-Sheng C, Wan-Ting T (2011) Failure analysis and risk management of a collapsed large wind turbine tower. Eng Failure Anal 18(1):295–313, 1350–6307. https://doi.org/10.1016/j.engfailanal.2010.09.008.

Kang H, Yugang Li, Fanghe Wu, Wie Guo, Caiyun Huan (2008) A system reliability analysis method for offshore wind turbine foundation. Electron J Geotech Eng 13

Sørensen J D, Toft H S (2010) Probabilistic design of wind turbines. Energies 3(2):241–257. https://doi.org/10.3390/en3020241

Yang M, Baniotopoulos C, Martinez-Vasquez P (2019) Wind turbine tower collapse cases: a historical overview. Proc Inst Civil Eng Struct Build 172(8):547–555.

Zafar U (2019) Probabilistic reliability analysis of wind turbines. Institute of Structural Mechanics, Bauhaus Universität Weimar, Masterthesis, Weimar

Zorzi G, Mankar A, Velarde Joey, Kirsch F (2020) Reliability analysis of offshore wind turbine foundations under lateral cyclic loading. https://wes.copernicus.org/articles/5/1521/2020/

Nuclear Power Plants

10

10.1 Introduction

So far, the discussion has taken place for the types of structures bridges, dams, tunnels, retaining structures, buildings, stadiums and wind turbines. Other types of structures are tanks and towers. However, no systematic studies are available for these at the moment. At this point it would be interesting to make a comparison with another technical product that does not belong to the class of structures. This would make it possible to check whether the comparisons made are sufficiently sensitive if systematic differences can be observed.

Theoretically, any technical product with a relevant safety risk can be used for this comparison. In fact, studies on the frequency and causes are available for a large number of such products, e.g. for the failure of tanks in chemical plants by Chang & Lin (2006). For aircraft, a whole phalanx of statistical evaluations exists, including complete aircraft loss. At the administrative level, for example, the European Union Aviation Safety Agency (EASA 2016), the European Transport Safety Council (ETSC 2003), DOT/FAA (2010), the International Air Transport Association (IATA 2018) and the International Civil Aviation Organisation (ICAO 2017) have published studies. At the commercial level, the major aircraft manufacturers produce such statistics (Boeing 2020; Airbus 2018).

In this book, however, nuclear power plants are to serve as the object of comparison. By far the most probabilistic calculations and statistical evaluations have been carried out for them.

D. Proske, *The Collapse Frequency of Structures*,
https://doi.org/10.1007/978-3-030-97247-9_10

10.2 Definitions

10.2.1 Definition of Nuclear Power Plants

According to the Swiss Confederation (Schweizer Eidgenossenschaft 2004), "*nuclear installations are ... facilities for the production of nuclear energy or for the extraction, processing, storage or rendering harmless of radioactive nuclear fuels and residues.* "

In Germany, the term nuclear facility is used instead of nuclear plants. According to the German Federal Ministry of Justice and Consumer Protection (Bundesministeriums der Justiz und für Verbraucherschutz 2020), a nuclear facility is a "*stationary installation for the production or for the treatment or processing or for the fission of nuclear fuels or for the reprocessing of irradiated nuclear fuels ... or serves the storage or interim storage of irradiated nuclear fuels*". In the context of this book, such types of plants are referred to as nuclear power plants if they are used to generate electricity.

10.2.2 Definition of Core Damage

Today, deterministic, and probabilistic safety formats are available for the safety assessment of nuclear power plants. These often, but not always, refer to a severe core damage. In probabilistic calculations, a distinction is made between three levels: Level-1, Level-2 and Level-3. In Level-1, the probability of a severe core damage or a severe fuel element damage is calculated. Level-2 includes the calculation of the probability of the release of radioactive material, and Level-3 includes the calculation of the probability of damage to health caused by these releases.

In this chapter, only probabilistic calculations Level-1, i.e. the calculation of the probability of a severe core damage, are evaluated. An example of the definition of severe core damage is the fulfilment of the following conditions:

- a maximum fuel cladding tube temperature of more than 1,200 °C,
- an oxidation of the cladding tubes of more than 17% of the cladding tube wall thickness,
- a maximum hydrogen production of more than 1% of the theoretically possible production and
- a change in the core geometry.

10.3 Stock of Nuclear Power Plants

Nuclear fission was first used in military technology during World War II. With the "Nuclear for Peace" programme, the knowledge was to be applied beneficially in the civilian sector in the Western world. However, the first nuclear power plant with a grid feed-in did not go into operation in the Western world, but in the Soviet Union in 1954. The first commercial nuclear power plant went into operation in Great Britain in 1956. Application on a large industrial scale began in the 1960s. In the early 1970s there were already 100 nuclear power plants worldwide, in the early 1980s about 250 and in the early 1990s around 400 nuclear power plants. Today, in 2021, there are about 450 nuclear power plants in operation worldwide (Statista 2020). The number of nuclear power plants has therefore not increased, or only slightly, over the last three decades.

Considering the reactors used for military purposes in nuclear-powered submarines and aircraft carriers, there are about 600 large-scale reactors in operation today. So-called zero-power reactors and research reactors were not counted. The results of probabilistic calculations used here refer to civilian reactors alone.

10.4 Calculation of Collapse Frequencies

10.4.1 Introduction

In contrast to structures, the scientific comparison of observed core damage probabilities and calculated core damage frequencies is the subject of numerous scientific publications. Such publications include WASH-1400 (1975), NUREG 1150 (1990), Kauermann & Küchenhoff (2011), Kaiser (2012), Lelieveld et al. (2012), Rangel & Leveque (2013), Wheatley et al. (2015), de Vasconcelos et al. (2015), Janke et al. (2016) and Raju (2016).

Figure 10.1 shows the core damage frequencies according to Kauermann & Küchenhoff (2011) and Wheatley et al. (2015). The figure also shows the results of a large number of probabilistic calculations.

10.4.2 Central Estimator

According to Prasser (2012), the core damage probability of the first-generation nuclear power plants was 10^{-3} to 10^{-4} per year. Own back-calculations with today's hazards show core damage probabilities of 10^{-3} per year for the nuclear power plants built at the end of the 1960s and 10^{-2} per year for the reactors in Fukushima. This also considers the increased core damage probabilities for multi-unit plants, which were neglected for a long time (Fleming 2013). Therefore, the worldwide calculated core damage probability in the 1970s was close to a value of one (per year between $100 \times 0.001 = 0.1$ and $100 \times 0.01 \approx 1$).

This rather high value can be confirmed both by the accidents in the 1970s and 1980s and by the numerous "near misses" (IAEA 2013). With the construction of new plants in the 1980s and 1990s and the improvement of existing plants in the 1990s, this value decreased. However, around the year 2000, there were still individual plants with

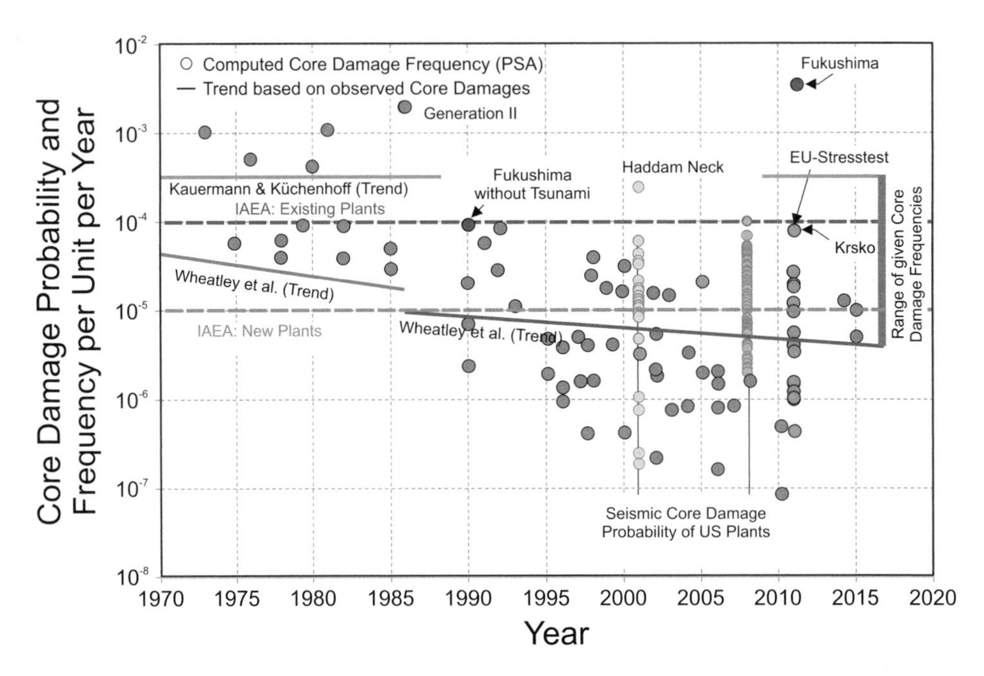

Fig. 10.1 Core damage frequencies and core damage probabilities for nuclear power plants (revised version from Proske 2016 in Proske 2020)

core damage probabilities between 10^{-3} and 10^{-4} per year, such as Haddam Neck or Fukushima (IAEA 2001). Today, target values are 10^{-4} per year for existing plants and 10^{-5} per year for new plants (IAEA 2001) are significantly undercut for new plants.

Nuclear power plants are subject to continuous monitoring and control by operators. Especially in the shutdown and low-power process, numerous operator actions are necessary in older plants. Therefore, human errors must be integrated into the probabilistic models in the form of human reliability analyses (HRA).

10.4.3 Measure of Deviation

The standard deviation of the probabilistic calculations and the observations is comparable. The frequency distribution of the two input variables also appears comparable. However, the calculation of the high core damage probability for Fukushima was retrospective and the calculation of the low core damage frequency refers to new plants (see Fig. 10.2).

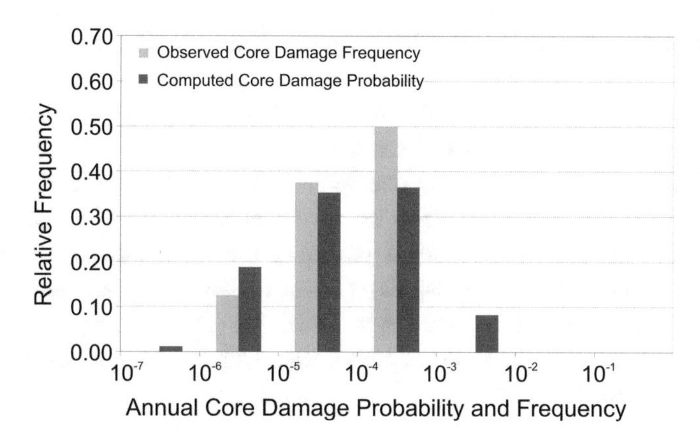

Fig. 10.2 Histogram of the calculated core damage probabilities and the observed core damage frequencies (Proske 2020)

10.4.4 Trend

Opinions differ widely on whether there is a falling trend in core damage frequencies. Kauermann & Küchenhoff (2011) do not see a falling trend, other authors, such as Rangel & Leveque (2013) and Wheatley et al. (2015) see a strong falling trend.

Assuming an exponentially decreasing trend, the decadic logarithm of the mean change per century is approx. 2.31 (Proske 2020). This corresponds to a reduction in core damage frequency by a factor of 200 extrapolated to a century. This is by far the largest value within the systems examined in this book. However, nuclear power plants are also the only group in which the mean observed value (core damage frequency) is greater than the mean calculated value (core damage probability).

10.5 Calculation of Failure Probabilities

The first probabilistic calculations of nuclear power plants date back at least to the work of Farmer in the late 1960s and the study WASH-1400 (1975) in the 1970tees. In Germany, the Risk Study of Nuclear Power Plants (Deutsche Risikostudie Kernkraftwerke 1980), commissioned by the Society for Reactor Safety (GRS), probably laid the foundation for probabilistic calculations. However, it was not until the mass application of the probabilistic calculation of core damage probabilities in the 1980s that the weaknesses of the first- and second-generation nuclear power plants became clearly visible.

For the evaluation here, the results of 85 probabilistic calculations of core damage probabilities were used (Fig. 10.1). Today, such calculations are available for practically all commercial nuclear power plants worldwide. Published results can be found, for example, in NUREG (2002), ENSREG (2012), Dedman (2011) and Mohrbach (2013).

As mentioned above, the IAEA (2001) target values are 10^{-4} per year for existing plants and 10^{-5} per year for new plants. These values may differ in individual countries. For nuclear power plants, for example, the Swiss Nuclear Energy Ordinance has target values for core damage probabilities for new buildings (Level-1). These also apply analogously to existing plants (KEV 2004 Article 24 Paragraph 1 Letter b, KEV Article 82, HSK 2008). Target values are also specified for Level-2 analyses (KEV 2004, Annex 3, HSK 2008). In a Level 3 analysis, the exposure of the population to radioactive material is calculated via the various transport paths. The latter studies are generally still of a research nature, whereas Level 1 and Level 2 analyses are to be prepared regularly and submitted to the authorities as proof of compliance with legal requirements (ENSI-A05 2009, ENSI-A06 2008).

10.6 Comparison

Figure 10.2 shows a comparison of the frequency distribution of the evaluation of core damage frequencies and core damage probabilities. The diagram includes the largest visual sample size of the probabilistic calculations compared to all structure types.

Nevertheless, Fig. 10.2 shows that both values agree relatively well on the shape of the distribution. As Table 10.1 shows, the standard deviations also agree well. The extreme values are more pronounced in the calculated core damage probabilities.

10.7 Causes of Collapse

With the probabilistic calculations for the nuclear power plants, systematic investigations of the causes of core damage are also available. These cause analyses are often used to select and plan plant improvements and reinforcements. Figure 10.3 shows the contributions to the core damage probability of a nuclear power plant. The total value is shown on the left. The remaining columns show the core damage probability for the respective initiating events. It can clearly be seen that earthquakes dominate the core damage

Table 10.1 Mean values including confidence intervals and standard deviations (Proske 2020)

Probabilistic calculations		Observed frequency		Target value per year
Standard deviation	Mean values incl. confidence interval	Standard deviation	Mean values incl. confidence interval	
3.35	2.08×10^{-5} (2.04×10^{-5}, 2.10×10^{-5})	3.41	2.37×10^{-5} (2.31×10^{-5}, 2.44×10^{-5})	10^{-5}

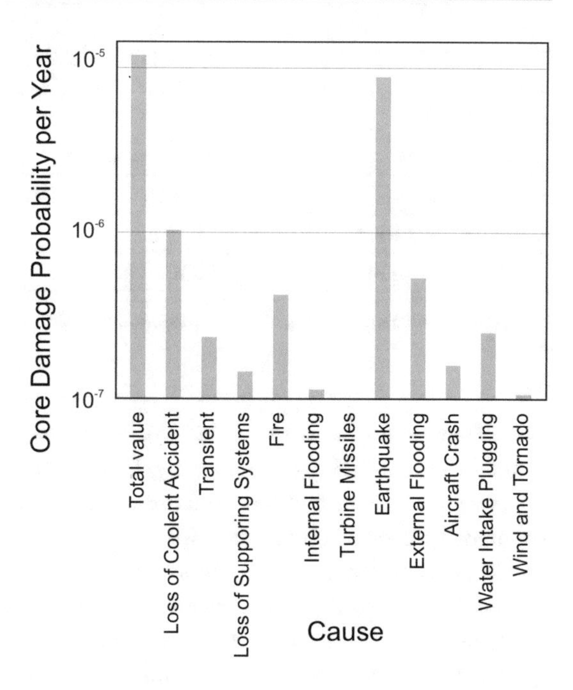

Fig. 10.3 Contributions of different causes to the calculated core damage probability

frequency in this case. If one wants to reduce the core damage frequency of this plant, one probably has to implement structural measures to increase the seismic capacity of the plant.

10.8 Mortality and Casualty Figures

In contrast to the determination of fatality figures in the case of structural failures, the determination of fatality figures because of severe accidents in nuclear power plants is associated with considerable uncertainties. One possible damage consequence of accident-related releases of radioactive material and associated ionizing radiation is deterministic and stochastic radiation damage. Stochastic radiation damage can lead to cancer as a long-term consequence. Due to the overall relatively large proportion of cancer as a cause of death in developed countries, a clear causal relationship cannot be established or is difficult to establish. For this reason, the fatality figures for the Chernobyl accident, for example, fluctuate by several orders of magnitude. Haury (2001), for example, cites a range in the literature from 42 to 500,000.

Determining fatality figures for building collapses is much easier here; the same applies to airplane crashes, for example.

Numerous studies compare fatality figures and mortalities in nuclear technology with other forms of electricity generation, such as Hauptmanns et al. (1987), Hirschberg et al. (1998), Inhaber (2004), ExternE (2007), Markandya & Wilkinson (2007), Burgherr &

Table 10.2 Various risk values for building and structures (Blockley 1980, Hauptmanns et al. 1987, Bea 1990, Cohen 1911, Paté-Cornell 1994, Menzies 1996, Proske 2009, Canisius 2015, Proske 2020)

Risk parameter	Buildings and structures[1]	Nuclear power plants[1]
Mortality per year	9.8×10^{-8}...1.5×10^{-6}	2.8×10^{-8} bis 5.8×10^{-7}
Fatal Accident Rate	0.002	0.0006
Loss of years of life	4.8 h	1.2 h

[1] Buildings and structures without earthquake and flood, nuclear power plants with earthquake and flood.

Hirschberg (2008), Preiss et al. (2013), Kharecha and Hansen (2013a, b), Ritchie (2017), Burgherr et al. (2019). Unfortunately, these results can only be used to a limited extent for comparison with building collapses, because here fatality figures are often related to the amount of electricity (joules, kWh, etc.) rather than in relation to a year. These units do not apply to structures. In the same way, mortality, and fatality figures for means of transport can be related to distance values. This also makes little sense for structures.

However, one can try to use other risk parameters, such as Fatal Accident Rates or "years of life lost". Such values are shown in Table 10.2.

10.9 Summary

Nuclear power plants are predestined for a comparison of collapse frequencies and failure probabilities because numerous and extensive studies of core damage frequencies and probabilities are available for them. In addition, there are numerous studies comparing the risks of all power generation technologies. Unfortunately, the results cannot be used directly for comparison with structures because different units are used.

The following observations can nevertheless be made:

- While the ratio of the mean calculated failure probability to the observed collapse frequency is greater than 1 for all types of structures - the calculations are therefore conservative - this is not the case for nuclear power plants. There seems to be a systematic difference here, which can possibly be explained by the different target values (10^{-4} for existing nuclear power plants and 10^{-5} for new plants) compared to buildings (10^{-6}). However, this is contradicted by the fact that the target value for nuclear power plants also includes consideration of human error, whereas this is excluded for structures. Another argument against this is that nuclear power plants are monitored almost permanently, which is not the case with structures.
- Systematic errors in the calculations can also be a problem, such as the simplified modelling of multi-unit plants (Fleming 2013). Individual errors, as shown by Epstein (2011) for Fukushima, will also be found in the calculations for structures. However,

due to the small absolute number of nuclear power plants and the absolutely few severe accidents, they can show significantly greater effects in the statistical parameters than for the approx. 1.5 billion structures.

- Apparently, operators, authorities and the public have become aware of the above-mentioned dilemma, because in none of the types of structures has such a rapid improvement in safety over time been observed as in nuclear power plants. The change value in relation to the century is much higher for nuclear power plants than for bridges, for example.

Based on these three arguments, it can be assumed that the original target values for the safety of nuclear power plants were chosen significantly too low. It can also be concluded that the approach used in this book may be sensitive enough for conclusions.

References

Airbus (2018) Commercial aviation accidents 1958–2017: a statistical analysis. Blagnac Cedex, France

Bea RG (1990) Reliability criteria for new and existing platforms. In: Proceedings of the 22nd offshore technology conference 7-10 May 1990, Houston, Texas, pp. 393–408

Boeing (2020) Statistical summary of commercial jet airplane accidents worldwide operations 1959–2019, boeing.com

Bundesministeriums der Justiz und für Verbraucherschutz (2020) Gesetz über die friedliche Verwendung der Kernenergie und den Schutz gegen ihre Gefahren (Atomgesetz) in der Fassung der Bekanntmachung vom 15. Juli 1985 (BGBl. I S. 1565), das zuletzt durch Artikel 3 des Gesetzes vom 7. Dezember 2020 (BGBl. I S. 2760) geändert

Burgherr P, Hirschberg S (2008) Comparative risk assessment of severe accidents in the energy sector. In: International Disaster and Risk Conference, IDRC, 25–29 August 2008, Davos

Burgherr P, Spada M, Kalinina A, Vandepaer L, Lustenberger P, Kim W (2019) Comparative risk assessment of accidents in the energy sector within different long-term scenarions and marginal electricity supply mixes. In: M Beer, E Zio (eds) Proceedings of the 29th European safety and reliability conference, Research Publishing, Singapore, , pp. 1525–1532. doi:https://doi.org/10.3850/978-11-2724-3_0674-cd

Canisius G (2015) Robustness of structures, October 2015, WSP/Parsons Brinckerhoff, Presentation, with Material from H Gulvanessian

Chang J, Lin C-C (2006) A study of storage tank accidents. J Loss Prev Proc Ind 19(1):51–59

Cohen BL (1991) Catalog of Risks extended and updated. Health Phys 61:317–335

de Vasconcelos V, Soares WA, da Costa ACL (2015) FN-Curves: Preliminary estimation of severe accident risks after Fukushima. In: 2015 International nuclear Atlantic conference (INAC 2015), Sao Paulo, Brasilien, 4–9 Oktober 2015, Associacao Brasileira De Energia Nuclear (ABEN), 12 p

Dedman B (2011) What are the odds? US nuke plants ranked by quake risk. NBCnews

Deutsche Risikostudie Kernkraftwerke (1980) Verlag TÜV Rheinland GmbH, Köln

DOT/FAA (2010) Trends in Accidents and Fatalities in Large Transport Aircraft, Final Re-port, US Department of Transportation/Federal Aviation Administration, DOT/FAA/AR-10/16, June 2010

EASA (2016) Annual Safety Review 2016, European Aviation Safety Agency, 2016
ENSI (2008) A06: Probabilistische Sicherheitsanalyse (PSA): Anwendungen, Richtlinien in Kraft: 13 Juni 2008
ENSI-A05/e (2009) Probabilistic Safety Analysis (PSA). Quality and Scope. Guidelines for Swiss Nuclear Installations. Swiss Federal Nuclear Safety Inspectorate ENSI, Edition March 2009
ENSREG (2012) EU Stress Tests and Follow-up, various documents. http://www.ensreg.eu/EU-Stress-Tests
Epstein W (2011) A Probabilistic risk assessment practitioner looks at the Great East Japan earthquake and tsunami. https://woody.com/wp-content/uploads/sites/35/2011/06/A-PRA-Practioner-looks-at-the-Great-East-Japan-Earthquake-and-Tsunami.pdf
ETSC (2003) Transport safety performance in the EU – a statistical overview. Brussel, European Transport Safety Council. http://etsc.eu/wp-content/uploads/2003_transport_safety_stats_eu_overview.pdf
ExternE (2007) Externalities of energy. http://www.externe.info/
Fleming KN (2013) Application of probabilistic risk assessment to multi-unit sites, Smirt 22. San Francisco
Hauptmanns U, Herttrich M, Werner W (1987) Technische Risiken. Springer Verlag, Berlin – Heidelberg GmbH
Haury H-J (2001) Die Zahl der Todesopfer von Tschernobyl in den deutschen Medien – ein Erklärungsversuch. GSF-Forschungszentrum für Umwelt und Gesundheit. April 2001
Hirschberg S, Spiekerman G, Dones R (1998) Project GaBE: Comprehensive assessment of energy systems: sever accidents in the energy sector, 1st edn, PSI-Bericht Nr. 98–16, November 1998
HSK (2008) Probabilistische Sicherheitsanalyse (PSA): Anwendungen, Ausgabe Mai 2008, Erläuterungsbericht zur Richtlinie A06/d
IAEA (2001) Safety Assessment and Verification for Nuclear Power Plants, Safety Guide, No. NS-G-1.2, International Atomic Energy Agency, Vienna
IAEA (2013) INES – The International nuclear and radiological event scale, user's manual. International Atomic Energy Agency, Vienna
IATA (2018) Safety Report 2017, Issued April 2018, International Air Transport Association, Montreal-Geneva
ICAO (2017) Safety Report, 2017 Edition, International Civil Aviation Organization, Montreal, Canada
Inhaber H (2004) Risk analysis applied to energy systems. Encyclopedia of Energy. Elsevier
Janke R, Stoll U, Grasnick C (2016) Nachfrage nach Nachrüstungen international, industrielle Trends; Rolle von Sicherheitsmargen, Nachrüstkonzepte und –möglichkeiten, atw, 61(2):116–124
Kaiser JC (2012) Empirical risk analysis of severe reactor accidents in nuclear power plants after Fukushima, science and technology of nuclear installations, Volume 2012, Article ID 384987, 6 p
Kauermann G, Küchenhoff H (2011) Reaktorsicherheit: Nach Fukushima stellt sich die Risikofrage neu, Frankfurter Allgemeine Zeitung, https://www.faz.net/aktuell/politik/energiepolitik/reaktorsicherheit-nach-fukushima-stellt-sich-die-risikofrage-neu-1605610.html. Accessed: 3 Dec 2019
KEV (2004) Kernenergieverordnung vom 10. Dezember 2004 (Stand am 1. Januar 2011), Der Schweizerische Bundesrat
Kharecha PA, Hansen JE (2013a) Prevented mortality and greenhouse gas emissions from historical and projected nuclear power. Environ Sci Technol 47(9):4889–4895
Kharecha PA, Hansen JE (2013b) Coal and gas are far more harmful than nuclear power. In: Global climate change: vital signs of the planet. NASA Goddard Space Flight Center

Lelieveld J, Kunkel D, Lawrence MG (2012) Global risk of radioactive fallout after major nuclear reactor accidents. Atmos Chem Phys 12:4245–4258

Markandya A, Wilkinson P (2007) Electricity generation and health. The Lancet 370(9591):979–990

Menzies JB (1996) Bridge failures, hazards and societal risk. In: International Symposium on the Safety of Bridges, July 1996. London

Blockley DI (1980) The nature of structural design and safety. Wiley & Sons, Chichester

Mohrbach L (2013) Fukushima two years after the tsunami – the consequences worldwide, atw, Vol. 58, Heft 3. März 2013:152–155

NUREG 1150 (1990) Severe accident risks: an assessment for five U.S. nuclear power plants, NRC, Washington, December 1990

NUREG 1742 (2002) Perspectives gained from the individual plant examination of external events (IPEEE) Program, NRC, Washington, April 2002

Paté-Cornell ME (1994) Quantitative safety goals for risk management of industrials facilities. Struct Saf 13:145–157

Prasser H-M (2012) Kernkraftwerke und Sicherheit, http://blogs.ethz.ch/math_phys_alumni/files/2012/11/Alumni_2012_11_13_Prasser.pdf

Preiss P, Wissel S, Fahl U, Friedrich R, Voß A (2013) Die Risiken der Kernenergie in Deutschland im Vergleich mit Risiken anderer Stromerzeugungstechnologien, Arbeitsbericht - Working Paper, Universität Stuttgart, Institut für Energiewirtschaft und Rationelle Energieanwendung, Bericht Nr. 11, Februar 2013

Proske D (2009) Catalogue of risks. Springer, Berlin-Heidelberg

Proske D (2016) Differences between probability of failure and probability of core damage. In: Caspeele R, Taerwe L, Proske D (eds) Proceedings of the 14th international probabilistic workshop, Ghent, Springer, pp. 109–122

Proske D (2020) Erweiterter Vergleich der Versagenswahrscheinlichkeit und -häufigkeit von Kernkraftwerken, Brücken, Dämmen und Tunneln. Bauingenieur 95(9):308–317

Raju S (2016) Estimating the frequency of nuclear accidents. Sci Glob Secur 24(1):37–62

Rangel LE & Leveque F (2013) How Fukushima-Daiichi core meltdown changed the probability of nuclear accidents? Saf Sci 64:90–98

Ritchie H (2017) It goes completely against what most believe, but out of all major energy sources, nuclear is the safest. https://ourworldindata.org/what-is-the-safest-form-of-energy

Schweizer Eidgenossenschaft (2004) Bundesgesetz über die friedliche Verwendung der Atomenergie (Atomgesetz), vom 23. Dezember 1959 (Stand am 27. Juli 2004)

Statista (2020) Number of operational nuclear reactors worldwide from 1954 to 2019. https://www.statista.com/statistics/263945/number-of-nuclear-power-plants-worldwide/

WASH-1400 (1975) Reactor safety study: an assessment of accident risk in the U.S. commercial nuclear power plants, NUREG 75/014, U.S. Nuclear Regulatory Commission, Springfield, October 1975

Wheatley S, Sovacool B, Sornette D (2015) Of disasters and dragon kings: A statistical analysis of nuclear power incidents & accidents. arXiv:1504.02380v1 [physics.soc-ph] 7 Apr 2015, 24 p

11 Concluding Remarks

11.1 Results and Critical Evaluation

In Chap. 1, the following theses were listed:

- Spaethe (1992): "*[the operational failure probability contains] only a share of the total failure probability … possible shares from human error [are] not included in this theoretical value. If the mechanical model can be assumed to be error-free, then the failure frequency will be greater than the theoretical failure probability.*"
- FWF (2018): "*theoretical probability of failure is orders of magnitude lower than the actual frequency of failures … actual failures are due to causes that are beyond theoretical probabilities …*"

Assuming that these theses are valid for the majority of buildings and structures, the investigations carried out here do not confirm these theses, whereby the first thesis also allows the conclusion that the mechanical models are all inaccurate, which of course, is true for all models.

Table 11.1 and Fig. 11.1 summarise the comparison for seven types of structures and clearly show that the average calculated failure probabilities in all cases are greater than the average observed collapse frequencies. However, the example of nuclear power plants in Table 11.1 shows that this is not necessarily so. For the nuclear power plants one can see that the calculated core damage probabilities are smaller than the observed core damage frequencies. The calculations are therefore not conservative, as desired, although human errors are explicitly taken into account in these probabilistic calculations, which should increase the calculated results.

Obviously, the need for action in nuclear engineering has been recognised, as shown by the large rate of change in Table 11.2 and Fig. 11.2. Figure 11.2 suggests a correlation

D. Proske, *The Collapse Frequency of Structures*,
https://doi.org/10.1007/978-3-030-97247-9_11

Table 11.1 Ratio of the mean calculated failure probabilities to the mean observed collapse frequencies (partly based on Hofmann et al. 2021)

I	II	III	IV
Type of Structure	Failure Probability	Collapse Frequency	Raw II to III
Bridge	2.00×10^{-4}	1.17×10^{-4}	1.71
Dam	3.46×10^{-4}	3.02×10^{-4}	1.15
Tunnel	5.30×10^{-4}	2.15×10^{-4}	2.47
Retaining Structure	1.35×10^{-3}	8.14×10^{-4}	1.66
Building	1.06×10^{-5}	3.30×10^{-6}	3.21
Stadium	2.80×10^{-4}	8.41×10^{-5}	3.33
Wind Turbine	1.23×10^{-4}	9.40×10^{-5}	1.31
Nuclear Power Plants	2.08×10^{-4}	2.37×10^{-5}	0.88

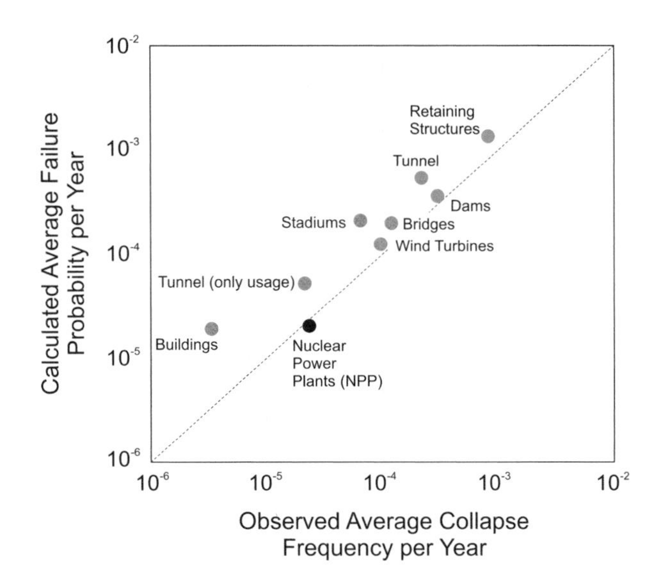

Fig. 11.1 Ratio of the mean calculated Failure Probabilities to the mean observed Collapse Frequencies

between the ratios of calculated values and observed values and the rate of change. Furthermore, Table 11.2 and Fig. 11.3 also seem to show a correlation between the year of the maximum observed values and the rate of change. In this context, we refer again to the study by Duffey & Saull (2003).

However, errors and weaknesses in the model can influence the results of the study and thus the conclusions. Such errors and weaknesses can be:

- The probabilistic calculations and the corresponding selected buildings and structures respectively may not be representative. This means that possibly only probabilistic

Table 11.2 Age, experience and change factors of the observed collapse frequencies for the different types of structures

Type of Structure	Age of Technology in Year	Worldwide Stock	Experience	Year of the maximum observed Collapse Frequency	Reduction of the Collapse Frequency per Century by Factor
Bridge	>2,000	5 … 6 million	2×10^7 Bridge Years since 1900 in the US	1850	$10^{0.67}$
Dam	>5,000	950,000	10^6 Dam Years since 1930 in Switzerland	1930	$10^{1.04}$
Tunnel	>4,000	125,000	-	2001	$10^{0.61}$
Retaining Structure	>6,000	26 million	-	-	-
Buildings	>10,000	1.5 billion	$>10^{11}$ Building Years	-	-
Stadiums	2,500	5,000	-	-	-
Wind Turbines	>4,000	450,000	-	-	-
Nuclear Power Plants	70	600	2×10^4 Reactor Years	1984	$10^{2.31}$

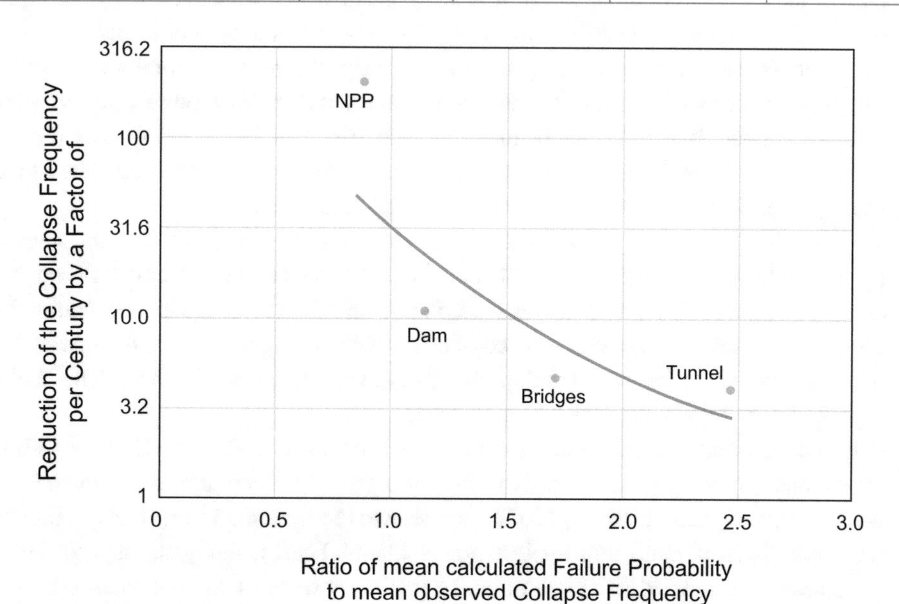

Fig. 11.2 Relationship between ratio of mean observed to calculated values and the rate of change of the observed values

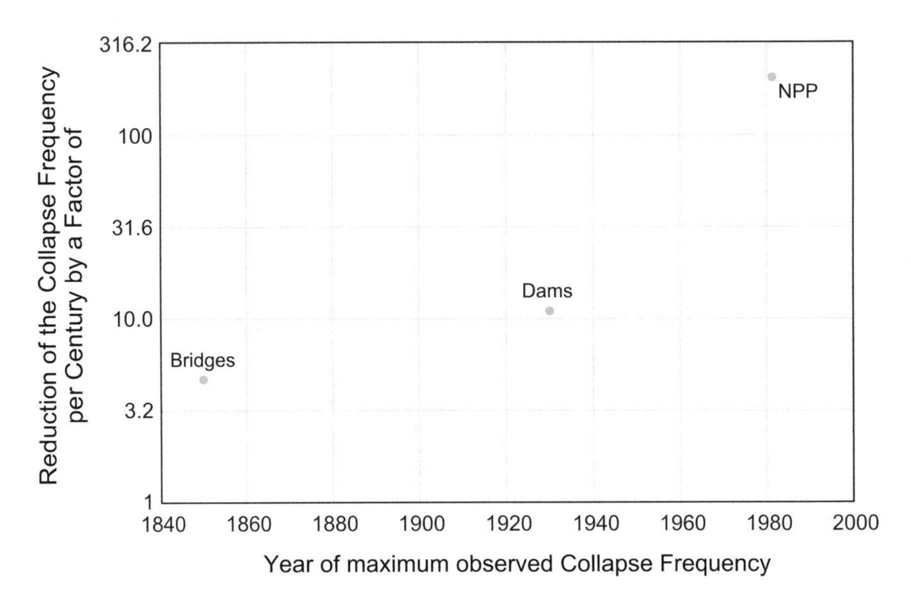

Fig. 11.3 Relationship between the year of maximum observed values and the rate of change of observed values

calculations of structures with a low or increased load-bearing capacity were used. In fact, at least for the bridges, the probabilistic calculations were selected to be representative of the bridge conditions in terms of the material and type of bridges.

- The probabilistic calculations may use qualitatively different structural and probabilistic models. For example, bridges can be represented as beam models or as 3D finite element models. Based on the evaluation of experimental load-bearing tests (Proske et al. 2021), it is assumed that this has a considerable impact on the calculated load-bearing capacities and thus also on the calculated failure probabilities.
- The sample sizes of the probabilistic calculations are too small. In fact, the number of probabilistic calculations is small compared to the number of collapses considered, see Table 11.3. Figures 11.4 and 11.5 show different results when evaluating the observed collapse frequencies and comparing with probabilistic calculations at different sample sizes for tunnel collapses. Further studies are needed here. This fact is especially true for stadiums and wind turbines.
- Selected probabilistic calculations were and are possibly flawed. Epstein (2011), for example, pointed out the errors in the probabilistic calculation of the Fukushima nuclear power plant. Fleming (2013) has shown the systematic error in probabilistic calculations of multi-unit nuclear power plants. Until a few years ago, the units were always calculated as single units. In reality, however, there are interdependencies between units in multi-unit plants, e.g. simply due to simultaneous effects such as

Table 11.3 Comparison of the number of probabilistic calculations and the statistical collapse frequency investigations considered (Proske 2020a, b)

Type of structure	Worldwide stock	Number of probabilistic calculations	Number of statistical collapse frequency investigations
Bridge	5 … 6 million	16	16[1]
Dam	950,000	25[2]	26
Tunnel	125,000	31	6 (+ 2[3])
Retaining Structure	26 million	7	11[4]
Buildings	1.5 billion	22	28
Stadiums	5,000	4	5
Wind Turbines	450,000	4	3
Nuclear Power Plants	600	85	9
Sum		194	106

[1] Some collapse frequency evaluations include more than 1,000 collapses
[2] Predominantly for seismic actions.
[3] Own statistical investigations of 321 collapses from two databases.
[4] Own statistical investigations for data after earthquakes.

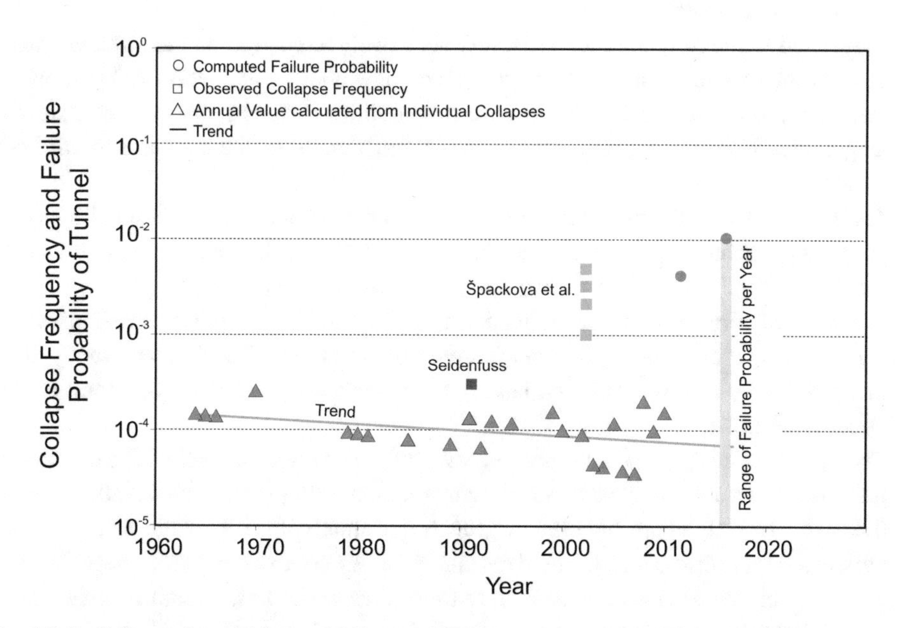

Fig. 11.4 Collapse frequencies and failure probabilities of tunnels (Proske et al. 2019)

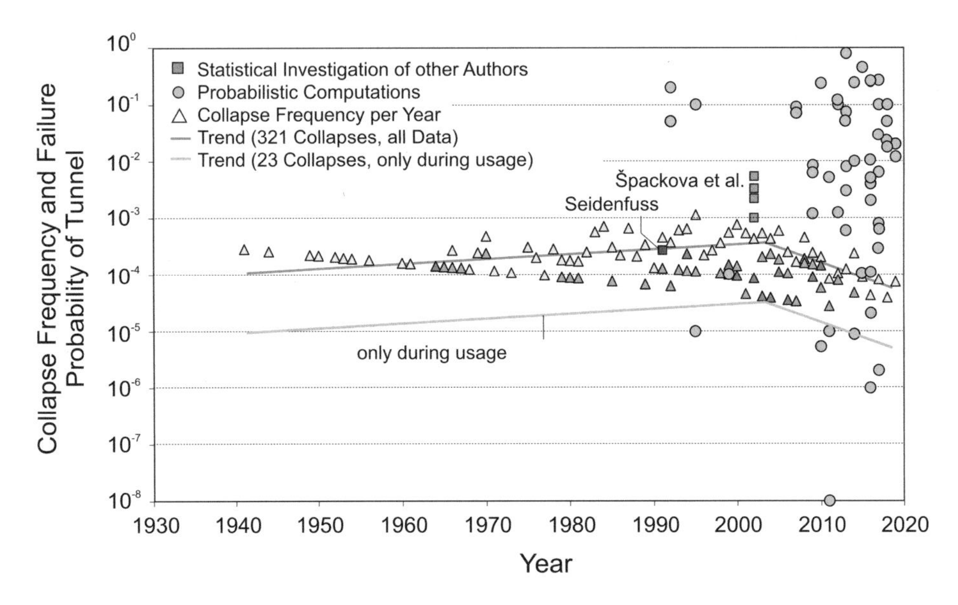

Fig. 11.5 Collapse frequencies and failure probabilities of tunnels (Spyridis & Proske 2021)

earthquakes or floods (common cause). This has significant effects on the calculated core damage probabilities.

- Calculated failure probabilities react very sensitively to changes in the input variables, such as the choice of distribution type. Oberguggenberger & Fellin (2005) show for one example that the calculated failure probabilities differ by up to eight orders of magnitude. Such considerable scatter affects the validity of the comparison with collapse frequencies.
- The comparison of the observed collapses and the calculated failure probabilities may not be admissible in this form. The differences between the two parameters and the limitation of the comparison have already been discussed in Chap. 1.
- In the calculation of collapse frequencies, flow quantities (number of collapses) are compared with inventory quantities (number of structures). The former usually show much larger fluctuations than the latter and the determination of the temporal anchor points is arbitrary.
- The collapse frequencies were only sporadically and not systematically calibrated to reference years and their population figures and building stocks. Especially in countries with very dynamic population growth and thus with a dynamic change of the building stock, differences to the determined values are to be expected here. In 2017 alone, 4.1 billion tonnes of cement were used worldwide, which should correspond to a considerable change in the number of buildings (Beckmann et al. 2021).
- It has already been pointed out in the chapter "Bridges" that the definition of the minimum bridge span differs in different countries. In addition, there are also structures

below the minimum bridge span. As a result, the real number of bridges may differ from the number used in this document. For example, in addition to the 6,000 bridges, the Swiss Railway (SBB) also manages a further approx. 3,500 culverts (Friedl 2019) that fall below the minimum bridge span. Such culverts also exist on other transport routes. Figure 11.6 shows a culvert on a hiking trail. If one were to multiply the number of bridges worldwide by the ratio of about 1.5 found for the SBB, one would obtain a worldwide inventory of 7.5 to 9 million bridges and culverts (see Table 11.2).

- Considering only observed collapses and neglecting observed damages leads to a limitation of the sample size and to an exclusion of observations that may be useful for the evaluation of the safety concept.
- Duffey & Saull (2003) do not use the calendar time for the rate of change, but the years of experience. This allows the number of structures and service life to be better taken into account than by the calendrical time, as in this contribution (Fig. 11.7). However, Duffey & Saull (2003) indirectly confirm another approach implemented in this book, as they compare not only dams and nuclear power plants, but a variety of industries. Thus, they also evaluate the rate of change for failures in medicine, drinking water supply or means of transport.

Fig. 11.6 Culvert on a hiking trail (Photo: *D. Proske*)

- Of course, it can be argued that the target values of failure probabilities are significantly lower than the observed collapse frequencies. However, this is a mixture of values for new and existing structures. In fact, age-dependent target values of failure probabilities are available, which also refute the conclusion (Proske 2018, see Fig. 11.8).
- Structures are designed for a specific service life. For example, Eurocode 0 (2017) or SIA 260 (2013) specify a scheduled service life of 50 years for buildings and 100 years for bridges. The maximum applicable service life can even be significantly longer, e.g. 200 years. However, the observed lifetimes and service lives can

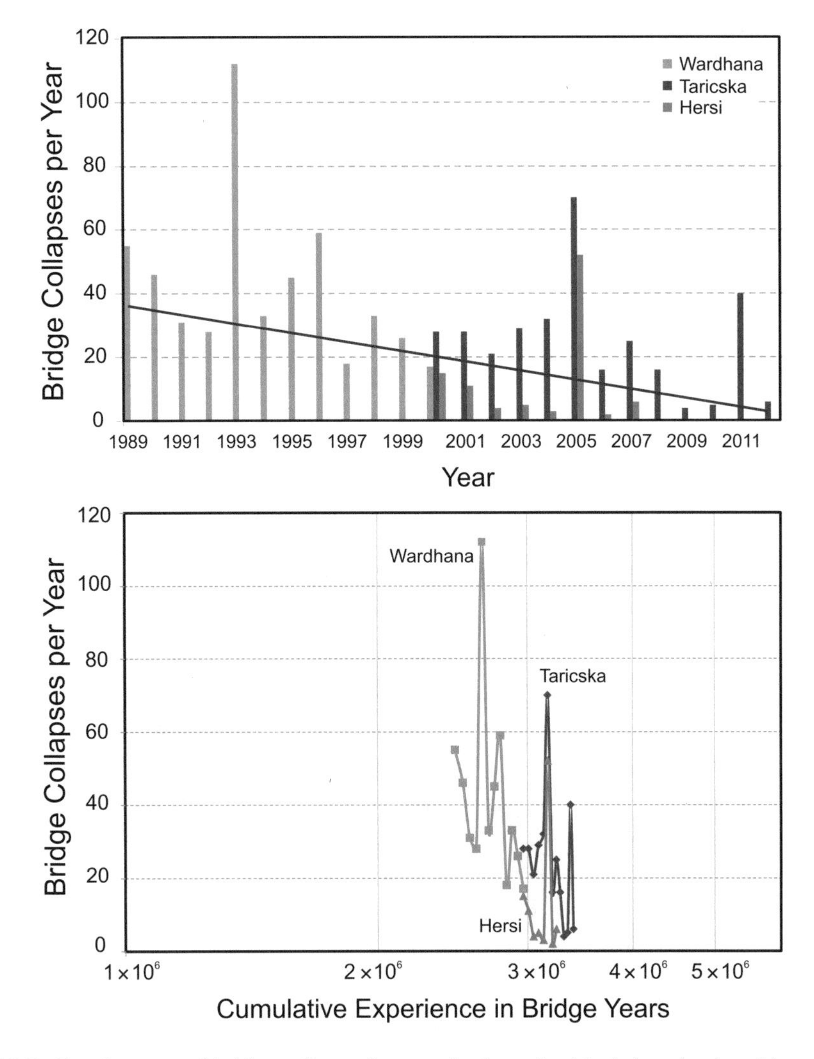

Fig. 11.7 Development of bridge collapse frequencies by calendrical time (top) and by years of experience (bottom)

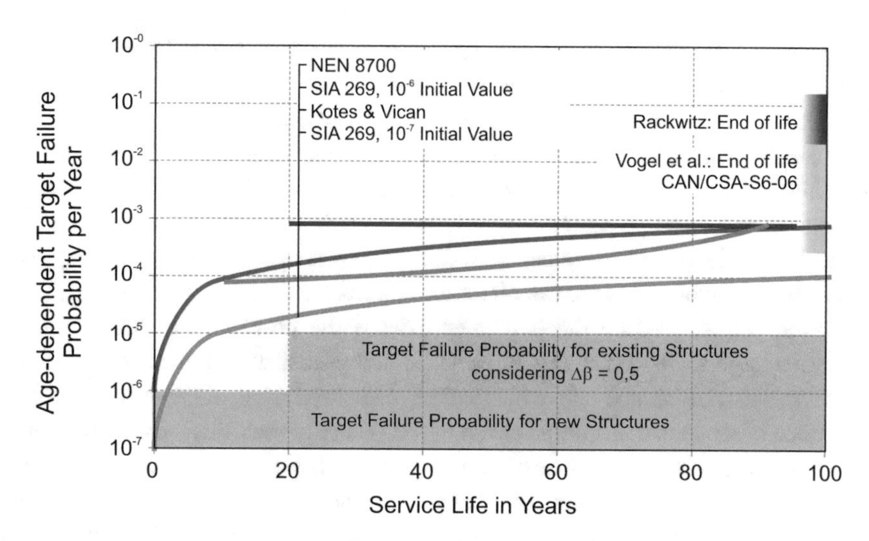

Fig. 11.8 Time-dependent target values of failure probabilities (Proske 2018; see also Kotes & Vican 2012 and Vogel et al. 2009)

deviate considerably from this. For example, reinforced concrete motorway bridges in the Netherlands have an average service life of 40 years (Klatter and van Noortwijk 2003), while railway bridges in Switzerland have an average service life of about 60 to 70 years. Some bridges are more than 2,000 years old (Proske & van Gelder 2009), some retaining structures more than 500 years old (Rück 2017). More than 50% of the buildings in Switzerland are more than 50 years old, and about 20% of the buildings were built before 1920 and are thus more than 100 years old (Gabathuler & Wüest 1989). And of course, many sacred buildings are hundreds of years old. This can have a significant impact on the calculated failure probabilities.

11.2 Added Value of Work

Comparing what scientists and engineers calculate with what they observe in reality is an essential part of scientific work (Popper 1993). Albert Einstein stated: "*All essential ideas in science are born out of the dramatic conflict between reality and our effort to understand that reality.*" The comparison between observation and computation is thus indispensable for the successful development and application of engineering and technology (Wallace 1971).

Kuhn (2012) has noted, however, that "*as long as the [scientific] tools provided by a paradigm prove capable of solving the problems it defines, science advances most rapidly and penetrates most deeply when those tools are used with full conviction. The reason is clear. As in manufacturing, so in science - a change of equipment is an extravagance that should be limited to absolutely necessary cases. The significance of crises lies in the indication they give that the time has come for such a change*." To check whether there is a crisis, one needs to know the observed values and critically examine the paradigms. This book represents a contribution to that discussion.

The use of probabilistic safety concepts for the vast majority of new buildings on earth, while arguing that the calculated failure probabilities should not be compared with collapse frequencies, as presented e.g. in Eurocode 0 (2017), raises the question of how to test the probabilistic safety concepts. In Chap. 1 it was discussed that failure probabilities and collapse frequencies are not directly comparable.

However, a link between the two must exist if the probabilistic safety concept is applied on a large scale to real structures and if collapses with administrative and political consequences influence the construction, structural analysis, reinforcement, and maintenance of structures. A simple check of the failure probabilities and their normative target values would be the identification of changes in the observed collapse frequencies due to new standards. In the context of this book, trend evaluations have been carried out that answer this question at least partly. Such comparisons are also carried out, for example, for damages caused by earthquakes in relation to the standards (see Fig. 11.9, but see also Bachmann 1997, 2002; Newson 2001).

Furthermore, Table 11.4 lists the estimated worldwide number of collapses for the different types of structures based on the collapse frequencies. These approximately 30,000 collapses do not consider collapses resulting from earthquakes and floods. In years with large-scale severe accidental actions, values exceeding 100,000 collapses can be observed (Table 11.5). An average value is 80,000 collapses per year due to earthquakes (Table 11.6). The number of collapses during wars and military conflicts is an order of magnitude greater.

The work is not finished, and further scientists will and should deal with the examination of probabilistic safety concepts. In this sense, this work does not represent a final, definitive assessment, but follows Pestalozzi's consideration mentioned here before "*It is the lot of men that no one has the truth. They all have it but distributed.*"

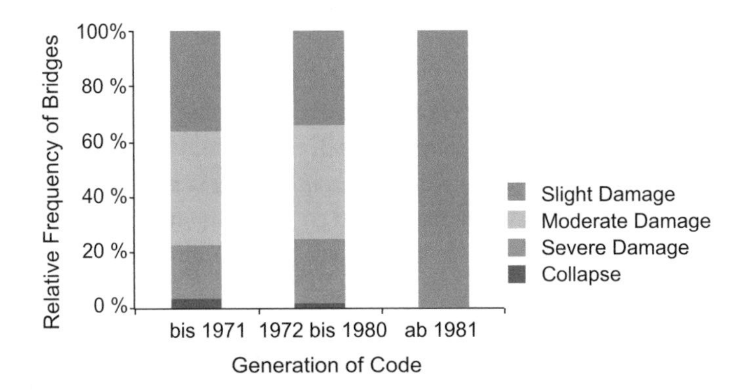

Fig. 11.9 Earthquake-induced damage to bridges in the Los Angeles area caused by the 1994 Northridge earthquake as a function of the generation of standards used for design (Wenk 2005)

Table 11.4 Global building stock and average number of collapses per year

Type of Structure	Worldwide Stock	Average number of Collapses per Year
Bridge	5 … 6 million	600
Dam	950,000	300
Tunnel	125,000	27
Retaining Structure	26 million	20,000
Buildings	1,5 billion	9,000[1]
Stadiums	5,000	0.40
Wind Turbines	450,000	42
Nuclear Power Plants	600	0.01

[1]without Earthquakes and Floods

Table 11.5 Examples of the number of buildings destroyed due to large-scale accidental actions

Year	Country	Action	Number of buildings destroyed
1970	Bangladesh	Flood	400,000
1995	Kobe, Japan	Earthquake	46,000–100,000
2001	El Salvador	Earthquake	200,000
2001	Peru	Earthquake	20,000
2004	Southeast Asia	Tsunami	300,000
2010	Haiti	Earthquake	250,000
2011	Japan	Earthquake and Tsunami	45,000–130,000 destroyed 190,000–240,000 damaged

Table 11.6 Worldwide number of damaged and destroyed buildings based on the Earthquake Impact Database

Year	Damaged buildings	Destroyed buildings	Worldwide collapse frequency
2015	660,000	85,000	6.5×10^{-5}
2016	380,000	90,000	6.9×10^{-5}
2017	560,000	115,000	8.8×10^{-5}
2018	450,000	95,000	7.3×10^{-5}
2019	340,000	90,000	6.9×10^{-5}
2020	170,000	30,000	2.3×10^{-5}

References

Bachmann H (1997) Erdbebensicherung der Bauwerke. In Mehlhorn G (ed) Der Ingenieurbau: Grundwissen, Teil 8: Tragwerkszuverlässigkeit, Einwirkungen. Verlag Wilhelm Ernst & Sohn, Berlin

Bachmann H (2002) Erdbebensicherung von Bauwerken, 2nd edn. Birkhäuser Verlag, Basel

Beckmann B, Bielak J, Scheerer S, Schmidt C, Hegger J, Curbach M (2021) Standortübergreifende Forschung zu Carbonbetonstrukturen im SFB/TRR 280. Bautechnik 98(3):232–242

Duffey RB, Saull JW (2003) Know the risk: Learning form errors and accidents: safety and risk in today's technology. Butterworth-Heinemann

Epstein W (2011) A probabilistic risk assessment practitioner looks at the Great East Japan Earthquake and Tsunami. https://woody.com/wp-content/uploads/sites/35/2011/06/A-PRA-Practioner-looks-at-the-Great-East-Japan-Earthquake-and-Tsunami.pdf

Eurocode 0 (2017) EN 1990 basis of structural design, 2nd edn, Draft 30 April 2017

Fleming KN (2013) Application of probabilistic risk assessment to multi-unit sites, Smirt 22. San Francisco

Friedl H (2019) Herausforderungen und Stossrichtungen im Anlagenmanagement für Brücken der SBB, Vortrag zum 1. Burgdorfer Brückenbautag mit Tagungsunterlagen, Burgdorf

FWF (2018) Austrian Science Fund, Expert Report: Comparison of probability of failure and frequency of collapse, 2. Juli 2018, Wien

Gabathuler C, Wüest H (1989) Bauwerk Schweiz: Grundlagen und Perspektiven zum Baumarkt der 90er Jahre, Zürich

Hofmann C, Proske D, Zeck K (2021) Vergleich der Einsturzhäufigkeit und Versagenswahrscheinlichkeit von Stützbauwerken, Bautechnik 98(7):475–481.

Klatter HE, van Noortwijk JM (2003) Life-cycle cost approach to bridge management in the Netherlands, transportation research circular E-C049: 9th International Bridge Management Conference, IBMC03–017, pp. 179–188

Kotes P, Vican J (2012) Reliability levels for existing bridges evaluation according to Eurocode, Procedia Engineering 40:211–216

Kuhn TS (2012) The structure of scientific revolutions, University of Chicago Press, 1962, 50th Anniversary Edition: 2012

Newson L ((2001) The Atlas of the world's worst natural disasters. Dorling Kindersley, London

Oberguggenberger M, Fellin W (2005) The fuzziness and sensitivity of failure probabilities. In: Fellin W, Lessmann H, Oberguggenberger M, Vieider R. (eds) Analysing uncertainty in civil engineering. Springer, Berlin, Heidelberg

Popper KR (1993) Alles Leben ist Problemlösen – Über Erkenntnis, Geschichte und Politik. Piper, München, Zürich

Proske D (2018) Bridge collapse frequencies versus failure probabilities. Springer

Proske D (2020a) Erweiterter Vergleich der Versagenswahrscheinlichkeit und -häufigkeit von Kernkraftwerken, Brücken. Dämmen und Tunneln. Bauingenieur 95(9):308–317

Proske D (2020b) Die globale Gesundheitsbelastung durch Bauwerksversagen. Bautechnik 97(4):233–242

Proske D, Spyridis P, Heinzelmann L (2019) Comparison of tunnel failure frequencies and failure probabilities. In: D Yurchenko, D Proske (eds) Proceedings of the 17th International Probabilistic Workshop, Edinburgh, pp. 177–182

Proske D, Sykora M, Gutermann M (2021) Verringerung der Versagenswahrscheinlichkeit von Brücken durch experimentelle Traglastversuche. Bautechnik 98(2):80–92

Proske D, van Gelder P (2009) Safety of historical Stone arch Bridges. Springer Verlag, Heidelberg

Rück Ph (2017) Natursteinmauern, Denkmal oder Risiko? Herbsttagung vom 26. Oktober 2017. Mitteilung Der Geotechnik Schweiz, Bern 175:89–96

SIA 260 (2013) Grundlagen der Projektierung von Tragwerken, Schweizerischer Ingenieur- und Architektenverein, Zürich

Spaethe G (1992) Die Sicherheit tragender Baukonstruktionen, zweite, neubearbeite Auflage. Springer, Wien

Spyridis P, Proske D (2021) Revised comparison of tunnel collapse frequencies and tunnel failure probabilities. ASCE-ASME J Risk Uncertainty Eng Syst, Part A: Civ Eng 7(2):04021004–1–04021004–9

Vogel T, Zwicky D, Joray D, Diggelmann M & Hoj NP (2009) Tragsicherheit der bestehenden Kunstbauten, Sicherheit des Verkehrsssystems Strasse und dessen Kunstbauten, Federal Roads Office (ASTRA), December 2009, Bern

Wallace WL (1971) The logic of science in sociology. Aldine de Gruyter

Wenk T (2005) Beurteilung der Erdbebensicherheit bestehender Strassenbrücken, Federal Roads Office (ASTRA), Bern

Printed in the United States
by Baker & Taylor Publisher Services